Microarrays in Diagnostics and Biomarker Development

Bertrand Jordan
Editor

Microarrays in Diagnostics and Biomarker Development

Current and Future Applications

Springer

Editor
Bertrand Jordan
CoReBio PACA
Luminy Science Park
13288 Marseille Cedex 9
France

ISBN 978-3-642-28202-7 ISBN 978-3-642-28203-4 (eBook)
DOI 10.1007/978-3-642-28203-4
Springer Heidelberg New York Dordrecht London

Library of Congress Control Number: 2012936972

Printed on acid-free paper

Springer is part of Springer Science+Business Media (www.springer.com)

Preface

This is the second time I assemble a book on microarrays for Springer—almost exactly 10 years after the first one (*DNA microarrays: Gene expression applications*. Springer Berlin Heidelberg 2001). During this interval, we have seen tremendous changes in the landscape of genomics and molecular biology. The first human genome sequence, with its revelations on gene number and relatively low human genetic diversity, was shortly followed by extensive investigation of single nucleotide polymorphisms in our DNA and, with the help of more and more complex "SNP arrays", by extensive whole-genome association studies (GWAS). These have uncovered many gene/phenotype correlations, but have also revealed how limited our understanding of the functional genome still is. The importance of non-coding transcripts and of epigenetic modifications of our DNA has been realised. And very recently, DNA sequencing technology has made tremendous progress, to the point that the "1,000-dollar genome" is in sight and that clinical use of whole-genome or exome sequencing is becoming significant (and is in fact competing with some uses of arrays). Meanwhile, clinical applications of DNA microarrays have developed, but not exactly in the directions anticipated 10 years ago and not quite at the level predicted by some analysts. Other microarrays using proteins, antibodies, peptide, or aptamer molecules as probes have made significant advances in spite of technical problems, and now have an impact not only in research but also in the clinic and in biomarker development.

This new book includes 13 main chapters, and covers essentially all types of arrays. It is focussed on entities that are in actual clinical use, or quite close to it—thus it does not discuss, for example, complex SNP arrays that remain essentially a research tool. It does present some very recent developments, such as peptide or aptamer arrays, or also miniaturisation towards "nanoarrays", that I see as having great potential in medicine even though their current presence is still limited. It discusses in detail very important issues in bioinformatics and in statistical analysis of array data, as well as the hurdles faced in the commercialisation of array-based tests and the vexing IP issues associated with these activities.

I believe this book will be useful to current array users who wish to have a complete view of the field, to newcomers who attempt to make the best choice between different technologies, and to academic scientists who engage in technology transfer activities and need to evaluate the hurdles involved in this process. I am especially grateful to the authors, my colleagues, who kindly accepted to undertake the absorbing task of writing a chapter, especially at a time where tight funding means that they are extremely busy preparing grant applications. And, of course, the organisational support from Springer to overcome the 1,001 small problems encountered in the production of this book is gratefully acknowledged.

Marseille, France Bertrand Jordan

Contents

Chapter 1
The Microarray Paradigm and Its Various Implementations

Bertrand Jordan

1.1 The Beginnings

The essence of the microarray approach is its "highly parallel" nature, that is, the fact that information on a large number of entities is obtained in a single experiment, thanks to a regular arrangement of "probes" (whatever their molecular nature may be) on some kind of support and on assessment of signals detected after incubation with a sample (usually labelled). The origins of this technology can be traced to the widely practised method of "colony screening" that was used, once the cloning revolution had begun in the early 1970s, to identify a few bacterial or phage clones of interest among the many thousands present in a given library (Fig. 1.1). It is true that the arrangement of clones in such experiments was random and therefore unrepeatable, but it was only one step from there to "ordered libraries" in which individual genomic or cDNA clones had been picked from Petri dishes, then distributed into microtiter plates, and could subsequently be arrayed on a suitable support (usually a nylon or nitrocellulose membrane). The essential feature was that detection of a positive (radioactive) signal at a given position on the membrane immediately identified specific clone(s) that could then be recovered from storage and used for further experiments (Fig. 1.2). This scheme was largely pioneered by Hans Lehrach's group in Germany, who adapted or developed the necessary instruments to perform these sophisticated (for the time) manipulations (Michiels et al. 1987; Craig et al. 1990; Lennon and Lehrach 1991). It was just a further step from there to use such "high-density filters" to assess gene expression, using arrays of cDNA clones (or their DNA) and labelled cDNA prepared from total mRNA of a given cell line or tissue (Gress et al. 1992), as well as the newly appeared imaging plate systems (Amemiya and Miyahara 1988) that allowed quantification of the signals. This approach was used by several groups to perform studies measuring simultaneously the

B. Jordan (✉)
CoReBio PACA, Luminy Science Park, 13288 Marseille, Cedex 9, France
e-mail: bertrand.jordan@univmed.fr

B. Jordan (ed.), *Microarrays in Diagnostics and Biomarker Development*,
DOI 10.1007/978-3-642-28203-4_1,

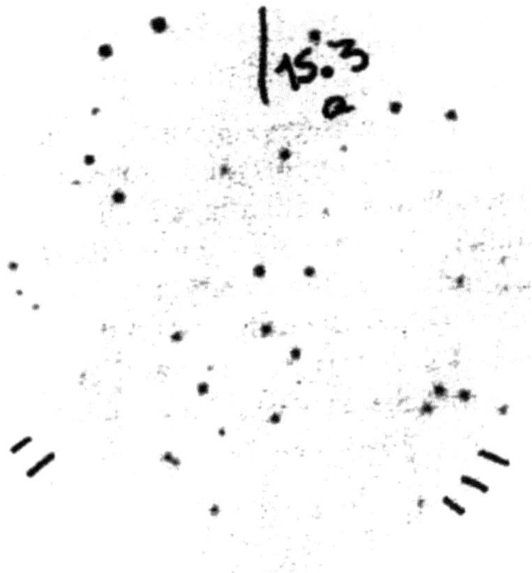

Fig. 1.1 A "colony filter" as used *circa* 1980 to find genes in clone libraries. Bacterial clones containing human DNA inserts are spread on nutrient agar in a Petri dish (ca 3,000 per dish). After growth of colonies, a nitrocellulose filter is placed on the plate, then removed and treated to lyse the bacteria and denature the DNA. The filter is then incubated with a radioactive probe corresponding to the gene of interest. After exposure to X-ray film, the "positive" clones can be seen and can be picked from the original Petri dish using the orientation marks drawn in radioactive ink

expression of hundreds or thousands of genes, an impressive feat at the time (Nguyen et al. 1995; Zhao et al. 1995) and, together with systematic tag sequencing of cDNA clones as pioneered by Craig Venter (Adams et al. 1991), gave the first general outlook on the human transcriptome as well as on that of a number of model organisms.

1.2 The Essential Miniaturisation Step

The detection system used in these experiments was radioactivity, that is, the cDNA prepared from the mRNA of the sample was labelled by incorporation of radioactive nucleotides containing 32P or 33P phosphate. Although this provided high sensitivity and a wide dynamic range, it severely limited the possible resolution and was also poorly adapted to many working environments. Thus, the first demonstration of an expression array using fluorescent labelling of the sample and detection with a confocal system (Schena et al. 1995; 1996) represented an important advance and is generally considered as the beginning of the microarray era. Meanwhile,

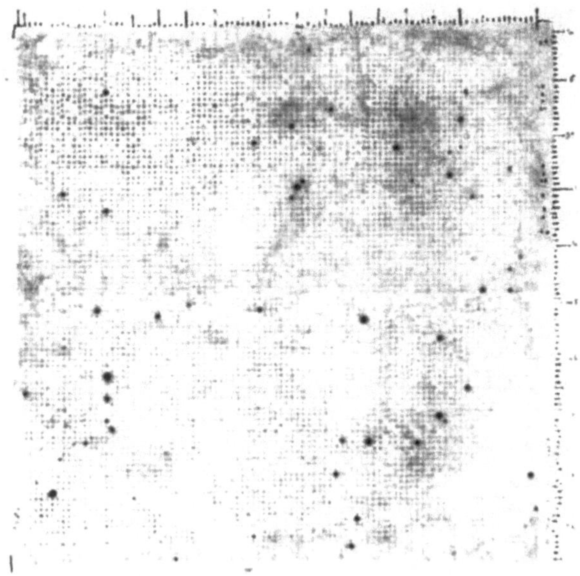

Fig. 1.2 A "high-density filter" used to access clone libraries (*circa* 1990). cDNA clones that have been picked from a library into microtiter plates are arrayed (100 × 100) on a large nylon filter. Hybridisation of this filter with a radioactive probe specific for the gene of interest detects several positive clones that can be directly recovered from storage for further study. Successive hybridisations of the same filter (or of replicas thereof) accumulate information on the 10,000 clones represented

the use, as probes, of oligonucleotides synthesised in situ had been described by Ed Southern as early as 1989 (see Fig. 1.3; Southern et al. 1992) and later by Steve Fodor (Pease et al. 1994). Rapid development followed, with commercial suppliers progressively taking over from research laboratories, and expression profiling using DNA arrays became a major approach in the late 1990s and early 2000s, with progressively more complex arrays that generally used oligonucleotides (pre-synthesised or synthesised in situ) as probes, and relied on fluorescent detection, although radioactivity kept some supporters and colorimetric methods were also implemented (Chen et al. 1998). By 2001, it had become possible to assess all (or most) of the human genes using a small set of commercial arrays, and by 2004, a single chip gave access of the whole human transcriptome.

1.3 Why Such Rapid Success?

The adoption of microarray technology (essentially for expression profiling in this first phase) was extremely rapid, both in terms of publications (hundreds in 2000, thousands in 2005) and of market size. It is clearly a technology that came at the right time: knowledge on genes and genomes had been accumulating for some time,

SYNTHESIS OF OLIGONUCLEOTIDES TETHERED TO A GLASS SURFACE: APPLICATIONS IN THE ANALYSIS OF NUCLEIC ACID SEQUENCES.
E M Southern and U Maskos; Department of Biochemistry, University of Oxford, U.K.

We have developed a new linker for oligonucleotide synthesis which leaves the oligonucleotide tethered to the glass surface after the deblocking step. The support bound oligonucleotide can be used for molecular hybridisation. The glass support can be in the form of microspheres, in which case synthesis and hybridisation can be conveniently carried out in columns.

We have also synthesised oligonucleotides in patches on the surface of microscope slides. This has the advantage that several different sequences can be synthesised on the same slide and the hybridisation signal detected by autoradiography. Applications to the analysis of nucleic acid sequences will be discussed.

Fig. 1.3 Abstract of the talk presented by Ed Southern at the 1989 "Genome mapping and sequencing" meeting at Cold Spring Harbor Laboratory (reproduced from the abstract book provided at the meeting). This is the first recorded example of oligonucleotide arrays, later published by the same authors (Southern et al. 1992)

and large-scale sequencing projects, after some initial difficulties, were getting on stream, one of the early landmarks being the sequence of yeast chromosome III (Oliver et al. 1992), all of 315 kb to be followed by much larger sequences in the following years, culminating with the first human sequence in 2001/2003. It was clear that there would be a pressing need to bring some functional significance to all these sequences, and expression profiling with microarrays appeared able to do so on the required scale. By the late 1990s, technology had advanced to the stage where this was feasible, with liquid handling and "spotting" robots to manufacture the arrays, and high-end confocal scanning devices for data acquisition. Informatics had also progressed, with the advent of the personal computer during the 1980s and of practical Internet at the beginning of the 1990s, so the required information processing and bioinformatics, while largely underestimated at the beginning, appeared feasible.

1.4 Generalisation of the DNA Array Approach

DNA arrays did not remain limited to expression profiling. Indeed, they had initially been thought of as sequencing devices (Khrapko et al. 1991; Southern et al. 1992), but this application did not succeed in providing a viable alternative to Sanger sequencing and was (provisionally) abandoned. However, another application progressively gained huge acceptance, the scoring of SNPs (single nucleotide polymorphisms or "snips") in human (or other) DNA. Large-scale genetic mapping, after the early eras of RFLPs and then of microsatellites, became based on the much more abundant snips (International HapMap Consortium 2005) and dependent on evermore complex oligonucleotide arrays that are able to assess first 10,000, then 500,000 and today several million snips in a person's DNA in a single experiment. These were intensively used in GWAS (genome-wide association studies) that revealed many genes influencing the risk for common, complex genetic diseases and also in DTC (direct to consumer) profiling of dubious predictive value but of strong commercial interest. In parallel, the realisation in the mid-2000s of the frequent occurrence in human DNA of insertions, deletions, duplications and inversions, collectively called CNVs, copy number variations (Redon et al. 2006), some of which have clinical relevance, motivated the development of specialised arrays (variably termed CGH arrays, CytoChips or CNV arrays) that today represent the major clinical use of DNA arrays.

1.5 Going Beyond DNA

The striking advantages of microarrays (parallel acquisition of many data points in controlled conditions and at moderate expense) did not escape the attention of scientists and clinicians in other fields, and extensive development efforts have been aimed for more than 10 years at protein, peptide, carbohydrate, tissue and even living cell arrays. Protein arrays in particular have been investigated in detail and under many different implementations (Joos and Bachmann 2009); they have great potential (and a number of actual uses) for clinical applications (Yu et al. 2010). Their development has nevertheless been relatively slow since many favourable features of nucleic acids are absent: procurement of the protein (antigen or antibody) "probes" to be spotted on the arrays is quite difficult (no easy synthesis as for oligonucleotides, no PCR for amplification), the protein/protein interaction on which detection is based is complex and very sensitive to many factors (unlike hybridisation of DNA and RNA, that is relatively well understood) and the data acquisition often requires complex two-step protocols ("sandwich" assays) that introduce further variability. Nevertheless, protein arrays are indeed being used in research and (usually in fairly low-plex form) in clinical diagnostics and can be expected to assume increasing importance. Peptide arrays have also been developed and successfully used, in particular for assays of kinase activities that have great importance in cancer drug development and use (Hilhorst et al. 2009). Carbohydrate arrays have also been reported (Chang et al. 2010) and could become important as glycosylation changes play a significant role in cancer. A particular and relatively low-tech application of arrays that has a major role in the clinic is represented by tissue arrays (Kononen et al. 1998). This consists of assembling several hundred slices of paraffin-fixed tissue from normal and pathological samples on a single glass slide. This slide can then be incubated with a given antibody, and the specificity of this reagent can be examined in detail (under the microscope) for each of the samples and correlated with the clinical information

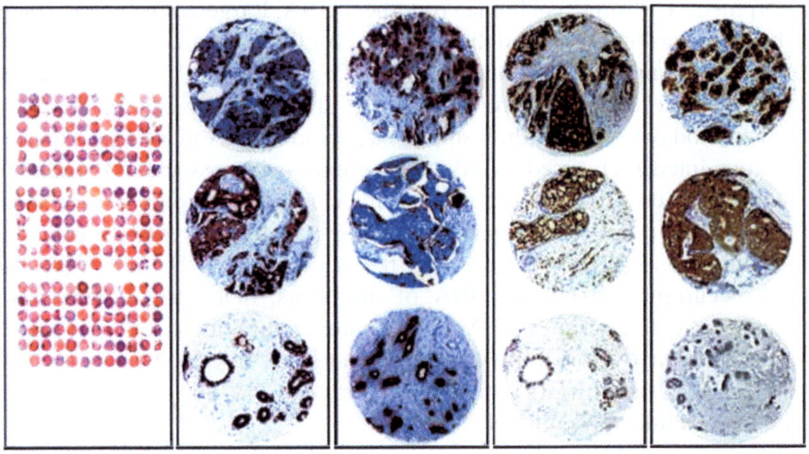

Fig. 1.4 Example of a tissue array. *Left*, general view of a microscope slide on which ~200 thin slices from as many pathological samples embedded in paraffin ("FFPE samples") have been assembled; *right*, blow-up of three of these slices after staining with four different antibodies. This allows precise characterisation of reagents on patient samples for which detailed clinical information is recorded

recorded. This simple application (Fig. 1.4) has great value and is a good example of the power of the microarray paradigm. Living cell arrays have also been developed for specialised applications such as large-scale RNAi (inhibitory RNA) screening (Wu et al. 2011).

1.6 Organisation of This Book

This book aims to provide an overview of current microarrays (DNA-based or otherwise), with emphasis on clinical and biomarker applications. Part I deals with DNA arrays in their various forms and uses; Part II with protein and peptide arrays, including also aptamer-based arrays that are protein-like in spite of their chemical structure; Part III with tissue arrays and miniaturised implementations of the microarray approach; and, finally, Part IV with data analysis and validation issues as well as commercial, regulatory and intellectual property aspects.

References

Adams MD, Kelley JM, Gocayne JD, Dubnick M, Polymeropoulos MH, Xiao H, Merril CR, Wu A, Olde B, Moreno RF et al (1991) Complementary DNA sequencing: expressed sequence tags and human genome project. Science 252:1651–1656

Amemiya Y, Miyahara J (1988) Imaging plate illuminates many fields. Nature 336:89–90

Chang SH, Han JL, Tseng SY, Lee HY, Lin CW, Lin YC, Jeng WY, Wang AH, Wu CY, Wong CH (2010) Glycan array on aluminum oxide-coated glass slides through phosphonate chemistry. J Am Chem Soc 132:13371–13380

Chen JJW, Wu R, Yang PC, Huang JY, Sher YP, Han MH, Kao WC, Lee PJ, Chiu TF, Chang F, Chu YW, Wu CW, Peck K (1998) Profiling expression patterns and isolating differentially expressed genes by cDNA microarray system with colorimetry detection. Genomics 51:313–324

Craig AG, Nizetic D, Hoheisel JD, Zehetner G, Lehrach H (1990) Ordering of cosmid clones covering the herpes simplex virus type I (HSV-I) genome: a test case for fingerprinting by hybridisation. Nucleic Acids Res 18:2653–2660

Gress TM, Hoheisel JD, Lennon GG, Zehetner G, Lehrach H (1992) Hybridization fingerprinting of high-density cDNA-library arrays with cDNA pools derived from whole tissues. Mamm Genome 3:609–661

Hilhorst R, Houkes L, van den Berg A, Ruijtenbeek R (2009) Peptide microarrays for detailed, high-throughput substrate identification, kinetic characterization, and inhibition studies on protein kinase A. Anal Biochem 387:150–161

International HapMap Consortium (2005) A haplotype map of the human genome. Nature 437:1299–1320

Joos T, Bachmann J (2009) Protein microarrays: potentials and limitations. Front Biosci 14:4376–4385

Khrapko KR, Lysov YuP, Khorlin AA, Ivanov IB, Yershov GM, Vasilenko SK, Florentiev VL, Mirzabekov AD (1991) A method for DNA sequencing by hybridization with oligonucleotide matrix. DNA Seq 1:375–388

Kononen J, Bubendorf L, Kallioniemi A, Bärlund M, Schraml P, Leighton S, Torhorst J, Mihatsch MJ, Sauter G, Kallioniemi OP (1998) Tissue microarrays for high-throughput molecular profiling of tumor specimens. Nat Med 4:844–847

Lennon GG, Lehrach H (1991) Hybridization analyses of arrayed cDNA libraries. Trends Genet 7:314–317

Michiels F, Craig AG, Zehetner G, Smith GP, Lehrach H (1987) Molecular approaches to genome analysis: a strategy for the construction of ordered overlapping clone libraries. Comput Appl Biosci 3:203–210

Nguyen C, Rocha D, Granjeaud S, Baldit M, Bernard K, Naquet P, Jordan BR (1995) Differential gene expression in the murine thymus assayed by quantitative hybridization of arrayed cDNA clones. Genomics 29:207–215

Oliver SG, van der Aart QJ, Agostoni-Carbone ML, Aigle M, Alberghina L, Alexandraki D, Antoine G, Anwar R, Ballesta JP, Benit P et al (1992) The complete DNA sequence of yeast chromosome III. Nature 357:38–46

Pease AC, Solas D, Sullivan EJ, Cronin MT, Holmes CP, Fodor SP (1994) Light-generated oligonucleotide arrays for rapid DNA sequence analysis. Proc Natl Acad Sci USA 91:5022–5026

Redon R, Ishikawa S, Fitch KR et al (2006) Global variation in copy number in the human genome. Nature 444:444–454

Schena M, Shalon D, Davis RW, Brown PO (1995) Quantitative monitoring of gene expression patterns with a complementary DNA microarray. Science 270:467–470

Schena M, Shalon D, Heller R, Chai A, Brown PO, Davis RW (1996) Parallel human genome analysis: microarray-based expression monitoring of 1000 genes. Proc Natl Acad Sci USA 93:10614–10619

Southern EM, Maskos U, Elder JK (1992) Analyzing and comparing nucleic acid sequences by hybridization to arrays of oligonucleotides: evaluation using experimental models. Genomics 13:1008–1017

Wu N, Castel D, Debily MA, Vigano MA, Alibert O, Mantovani R, Iljin K, Romeo PH, Gidrol X (2011) Large scale RNAi screen reveals that the inhibitor of DNA binding 2 (ID2) protein is

repressed by p53 family member p63 and functions in human keratinocyte differentiation. J Biol Chem 286:20870–20879
Yu X, Schneiderhan-Marra N, Joos TO (2010) Protein microarrays for personalized medicine. Clin Chem 56:376–387
Zhao N, Hashida H, Takahashi N, Misumi Y, Sakaki Y (1995) High-density cDNA filter analysis: a novel approach for large-scale, quantitative analysis of gene expression. Gene 156:207–213

Part I
DNA Arrays

Chapter 2
DNA Arrays in Many Guises

Bertrand Jordan

2.1 The Birth and Development of DNA Arrays

The first generally recognised publication on DNA microarrays appeared in 1995 (Schena et al. 1995), although a number of DNA array studies had been published previously, generally using nitrocellulose or nylon supports and radioactive labelling of the sample, options that were not compatible with extensive miniaturisation (Lennon and Lehrach 1991; Gress et al. 1992; Zhao et al. 1995; Nguyen et al. 1995). In-house production of microarrays was widespread in the first years of the technology (DeRisi and Iyer 1999), but industry gradually took over and provided great advances in quality control and, ultimately, real cost. The DNA "probes" used to construct arrays were initially amplification products from cDNA clones, but these were gradually superseded by oligonucleotides: management of large cDNA libraries is very cumbersome and error-prone and, contrary to some expectations, relatively short oligonucleotides turned out to provide satisfactory specificity on hybridisation. Currently, essentially all DNA microarrays consist of oligonucleotides arrayed on a solid support, usually glass, but also occasionally plastic or silicon. Where they differ is in the length of oligonucleotides. These may be short (20–25 bases) and synthesised in situ, as for Affymetrix, in which case a number of overlapping sequences are used to assess the expression of every gene, or long (50–70 bases), synthesised in situ (Agilent, Roche Nimblegen) or pre-synthesised (Illumina), in which case a single oligonucleotide may be sufficient to measure the expression of one gene. Each of the four major approaches has its strong and weak points. The microelectronics-like light-directed synthesis process developed at Affymetrix allows the efficient production of extremely complex chips but is relatively inflexible as any change in content involves expensive retooling (manufacture of a new set of very high precision masks); the Nimblegen process is also based on light-directed synthesis but uses a programmable digital

B. Jordan (✉)
CoReBio PACA, Luminy Science Park, 13288 Marseille, Cedex 9, France
e-mail: bertrand.jordan@univmed.fr

B. Jordan (ed.), *Microarrays in Diagnostics and Biomarker Development*, 11
DOI 10.1007/978-3-642-28203-4_2, © Springer-Verlag Berlin Heidelberg 2012

micromirror device[1] instead of masks, which makes it much more flexible. Agilent's in situ manufacturing system, based on successive delivery of synthesis reagents to each spot, is readily programmable and extremely flexible, well adapted to production of custom arrays, but complexity is somewhat lower and economies of scale less evident. Finally, the Illumina process relies on pre-synthesised long oligonucleotides and is extremely miniaturised but, again, not very flexible (see Table 2.1).

Progress in miniaturisation has resulted in arrays that can monitor the expression of all known human genes or score up to one million single nucleotide polymorphisms. Spot or "feature" size is now typically below 50 μm and can be as low as a few micrometres. In research applications, the sample is labelled with a fluorophore (usually Cy3 and/or Cy5), and data acquisition is achieved by laser scanning and confocal detection. This provides for very high resolution and sensitivity, allowing the use of complex arrays, but requires expensive apparatus and relatively cumbersome procedures. For clinical purposes, where the array is often less complex, labelling may involve a fluorophore, but also chemiluminescence or even colorimetry, and detection with a very affordable CCD unit may be adequate. In some cases, electrical detection systems allow the assay of unlabelled samples. In terms of cost, expression arrays covering the whole human transcriptome can now be obtained (in quantity) for as little as €100, down from several thousand at the time of their introduction in the early 2000s.[2] Diagnostic applications often call for moderately complex arrays, but the cost of an array is now more dependent on the size of the production run than on its complexity; thus, relatively simple custom arrays may be as expensive as highly complex standard chips. Of course, in diagnostics, cost issues are essential; thus, the compromises made and the choice of manufacturer can be quite different from those that apply in research. In most cases, manufacturers now provide their products as multi-array slides (e.g. 12 arrays of 135,000 oligonucleotides each for Nimblegen), a format that helps to bring down the per-array cost and helps with reproducibility issues but introduces some operational constraints.

2.2 Trends in Research

During the initial years, DNA arrays were essentially used for expression studies, and expression profiling became one of the major exercises in large-scale genome analysis, especially as, at that time, sequencing technology was still based on the Sanger technique, very slow and expensive by present-day standards. Of course, compared to previous methods (essentially Northern blotting), microarrays were incredibly powerful, and accordingly, their uptake by the research community was

[1] Similar to the devices found in video projectors.

[2] In March 2001, the list price, in Europe, for a set of five Affymetrix arrays (U95 A to E) representing the whole human transcriptome (as far as it was known at the time) was €6,400.

Table 2.1 Four major approaches to DNA array manufacture

Company	Outline of method	Typical feature size (μM)	Oligonucleotide length	Number of oligonucleotides per gene (expression)	Number of features per array	Cost for large production runs	Flexibility
Affymetrix	In situ light-directed synthesis (mask lithography) of short oligonucleotides	10	15–20 nt	10–20	Up to 1 M	Very low	Limited (retooling necessary)
Roche Nimblegen	In situ light-directed synthesis (programmable micromirror device) of long oligonucleotides	13	60 nt	3	Up to 4 M	Low	Good (reprogramming of micromirror device)
Agilent	In situ synthesis of long oligonucleotides by programmed reagent delivery	50	60–70 nt	1–2	Up to 200 K	Low	Excellent (reprogramming, custom order on Internet)
Illumina	Pre-synthesised long oligonucleotides deposited in several microwells	3	60–70 nt	~100 identical in different wells	Up to 4 M	Very low	Relatively limited

This table shows the salient features of the approach adopted by four major manufacturers of arrays. Given the wide variety of arrays produced by these companies, the figures given can only be an approximation

http://media.affymetrix.com/support/technical/other/clinical_toolkit_product_spotlight.pdf

http://www.nimblegen.com/company/technology/

http://www.genomics.agilent.com/

http://www.illumina.com/applications/

extremely rapid: the number of publications on expression profiling rose from a few in 1995 to several hundred in 2000 and nearly 2,000 in 2005 (Lenoir and Giannella 2006). It quickly became apparent, however, that much of the early data was of dubious validity because of technical issues with the platforms used, insufficient use of replicates and over-optimistic interpretation. Serious as they are for research projects, these problems were an even stronger deterrent for clinical applications. However, with growing realisation of the importance of these issues (Tan et al. 2003; Jordan 2004), these problems were tackled and gradually resolved, thanks to industry standardisation of arrays and to the efforts of the FDA-sponsored MAQC Consortium (2006). Currently, however, profiling methods based on sequencing (RNA-seq) are becoming dominant in research applications, as they provide higher quality data (digital rather than analogue), with information on mutations and on "new" transcripts, at a cost that is now approaching that of an expression array experiment.

Meanwhile, the development of DNA arrays capable of scoring large numbers of single nucleotide polymorphisms (SNPs or snips) made great progress, and by the early 2000s, chips able to assess 10,000 snips in a single experiment were available and became widely used (Gunderson et al. 2006). Current snip arrays can now score 1,000,000 or more of these markers and have been essential for whole-genome association studies in which several thousands of patients and controls are genotyped to discover significant associations between loci in the genome and vulnerability to common diseases such as diabetes, Crohn's disease or even autism (Lam et al. 2010). While these studies have their problems, they do provide valid and useful results and have been extensively practised, providing a large market for these snip arrays that are now the dominant species in the DNA array world. Snip arrays are also used for direct to consumer (DTC) genetic profiles whose predictive value is however extremely limited.

Beyond snip genotyping, the growing recognition of the existence of numerous deletions, duplications and insertions (collectively called structural variations) in our DNA, some of which display clear clinical correlations, has motivated the development of arrays capable of detecting these events. Suitable probes may be included in snip arrays, or, alternatively, specialised "CGH arrays" may be developed to probe the whole genome for structural variations at various levels of resolution. This has become an important application of DNA arrays in research and, indeed, also in the clinic (see Chap. 5).

In summary, research use of DNA arrays has largely shifted from expression profiling to snip scoring and structural variation studies but remains very significant in spite of strong competition by next-generation sequencing for some applications.

2.3 Great Expectations for Clinical DNA Arrays

With the personalisation of medicine, made possible by our knowledge of the human genome and our growing, if still limited (Ashley et al. 2010), understanding of its influence on pathological conditions, molecular diagnostics are taking an increasingly prominent role. The size of the molecular diagnostic market worldwide was estimated at 3–4 billion USD in 2010 (Medical News Today 2008), and it is expected to grow at an annual rate of 20%. While this figure is dwarfed by those of medical drugs, where cancer medicines alone account for 50 billion USD (Chapman 2010), diagnostics are nevertheless an increasingly essential part of medical care and are attractive from a business point of view since development schedules, regulatory issues and costs are generally more favourable than for drugs. Apart from clinical diagnostics *stricto sensu*, the drug development process relies more and more on various types of biomarkers to provide early assessment of toxicity or off-target effects of potential drugs and, increasingly, on selection for clinical trials of those patients more likely to respond to the molecule being tested by virtue of their particular genetic makeup. DNA arrays, in their various implementations, obviously have an important role to play in these developments.

Indeed, as soon as they appeared, DNA arrays were hailed as potentially revolutionary diagnostic devices. Schena et al. (1995) stated that "microarrays of cDNAs could thus provide a useful link between human gene sequences and clinical medicine". In the early 2000s, clinical microarrays were widely expected to represent a multibillion (USD) market in the near future (Moser 2001). DNA arrays have indeed been adapted to clinical use and, in some cases, have obtained FDA clearance in the USA: the most widely publicised examples are the CYP450 AmpliChip, developed by Affymetrix and Roche and aimed at characterising the genes for two enzymes involved in drug metabolism (Jain 2005), and the MammaPrint microarray test, proposed by the Dutch company Agendia and designed to analyse expression profiles in resected breast tumour material to indicate the likelihood of tumour recurrence (Glas et al. 2006). Both eventually obtained FDA clearance and were followed by several other array-based tests.

Nevertheless, the impact of microarrays in molecular diagnostics has remained relatively limited (Li et al. 2008; Jordan 2010): their market is currently estimated to be 300–400 million USD, i.e. less than 10% of the total molecular diagnostics market. In many discussions emphasising the rosy perspectives of clinical DNA arrays, important requisites for a diagnostic test were lost from sight, and the differences between the research environment and the clinical laboratory were sometimes overlooked. A diagnostic test must be very robust technically, even when performed under suboptimal conditions and by moderately qualified personnel. It must also provide a clear-cut result, a clear "yes", a definite "no" or a well-defined numerical score on which therapeutic decisions may be based. Regulatory approval is also essential. While this is a relatively minor issue for research equipment and products, it becomes an absolute requirement for a clinical test that aims to be eventually accepted for reimbursement by public or private health maintenance organisations. Various agencies and regulations are involved, and they may heavily influence technological choices, usually in favour of methods that are considered "established" (see Chap. 12).

The most important necessity, however, is real-world clinical utility: the test must provide solid input into therapeutic decisions, and its use must be shown to improve the outcome for the patient (Børresen-Dale 2003; Simon 2008). To take a widely discussed example, a major issue in cancer management is the avoidance of unnecessary chemotherapy. Only a fraction of tumours turn out to have metastatic potential, and only a subset of those respond to treatment. A variety of clinical indexes based on tumour size, histochemical examination and clinical presentation have been developed, but the information they provide is insufficient, and as a result, many patients are subjected to an unpleasant, dangerous and expensive therapy that is actually useful only for a fraction of them. Expression profiling performed using a fragment of tumour tissue can provide indications on the aggressiveness of the cancer and thus help orient treatment but will be of real use only if the result allows, for example, definite identification of a class of patients whose risk is so low that additional therapy can be safely withheld. Otherwise, the test will neither be widely performed nor reimbursed by insurance companies, an essential issue given that its cost is usually quite high.[3]

2.4 Expression Arrays Versus Genotyping Arrays

The distinction formulated in the subtitle above is essential, for both technical and for regulatory reasons (see Table 2.2). *Expression arrays*, possibly the most familiar use of DNA arrays and historically the initial one, measure the expression level of a set of genes in a tumour, tissue or blood sample from a patient. *Genotyping DNA arrays,* on the other hand, are aimed at characterising a DNA (or, in some cases, RNA) sequence present in the sample. They may be designed for bacterial or viral identification or, in the case of human DNA, to find out if specific mutations are present, to define alleles for a number of sites where single nucleotide polymorphisms (snips) have been reported or to characterise copy number variation.

These two applications are technically quite dissimilar; they involve different sample handling procedures and entail diverse requirements on storage conditions since mRNA is much more labile than DNA. However, the most important difference between them lies in the solidity of clinical correlates. To use expression profiling in the clinic, the company introducing the test relies on new studies (performed in-house or with academic collaborators) showing that a given expression profile is prognostic (and possibly predictive[4] as well). While these studies have usually been published in reputable journals, they may not be generally accepted, and even if they are, the issue of clinical utility remains: it is up to the company (and the collaborating academic group) to show that using the test will measurably improve the outcome for patients. Indeed, for some of the major expression tests currently on the market, large clinical studies are still under way, and their final results will not be available for several years. Under these conditions, obtaining FDA clearance for clinical use, gaining physician acceptance and achieving

[3] The breast cancer prognostic tests discussed in Chap. 3 all cost at least €2,000, worthwhile only if their use can lead to avoidance of unnecessary chemotherapy.

[4] For the precise meaning of "prognostic" and "predictive", see Chaps. 3 and 11.

Table 2.2 Expression versus genotyping applications for DNA arrays

Application	Complexity of arrays	Importance of sample quality	Sensitivity requirements	Clinical utility of result	Acceptability as LDT ("home brew")	Potential for regulatory approval
Expression profiling	Low to high	Critical (fresh frozen tissue in most cases)	High (confocal scanning)	Generally to be proven	Disputed by FDA	Possible but complex (IVDMIA)
DNA genotyping	Low to high	Not critical (within limits)	Moderate (CCD detection or other methods possible)	Usually known from previous research and clinical experience	Widely accepted, so far	Relatively good

This table shows the different requirements and characteristics of these two types of application
LDT Laboratory Developed Test; *IVDMIA* In Vitro Diagnostic Multivariate Index Assay

reimbursement from health maintenance organisations are all quite difficult and limit the market penetration of these tests that carry a typical cost of approximately €2,000.

For genotyping applications, on the other hand, the clinical significance of the molecular data is generally known in advance. This is true for bacterial or viral identification: for example, previous knowledge indicates whether or not the detected HPV (human papillomavirus) subtype in a cervical sample corresponds to a low or high risk of subsequently developing cancer. This also holds for mutations in genes known to be involved in disease: the pathological consequences of a given mutation in the BRCA1 gene, for example, are usually well known. These correlations are widely accepted both by clinicians and by regulatory agencies. Thus, as long as the diagnostic test is technically correct, the clinical significance of the result is clear, and, accordingly, regulatory approval for this category of test is relatively straightforward, and acceptance is essentially a matter of reproducibility, ease of operation and cost. For genome-wide scanning of copy number changes the situation is less clear, as the field is still evolving rapidly, but a number of deletions and duplications now have well-established pathological correlates. Finally, whole-genome snip scoring is still very much a research tool and does not at this time have applications for clinical diagnostics except for "recreational" direct to consumer (DTC) testing of very dubious medical value. Apart from this last case, it remains true that diagnostic tests that examine the genetic makeup of a microorganism, a tumour cell or a human patient can build on solid knowledge of the clinical effects resulting from DNA sequences and thus enjoy a relatively smooth path to regulatory clearance, acceptance and reimbursement.

2.5 Outline of the Following Chapters

As could be expected, a sizeable fraction of this book is devoted to DNA arrays. The next two chapters describe in some detail clinical applications of expression and genotyping arrays, with several examples of diagnostic tests that are currently in use. The following one (Chap. 5) is dedicated to "CGH arrays", increasingly used in clinical diagnostics to complement or replace conventional karyotypes and FISH (fluorescence in situ hybridisation) analysis. Further along in the book, after chapters that deal with other types of arrays (notably protein, peptide and tissue arrays), several general issues that impinge on all types of arrays are discussed: miniaturisation (nanoarrays), data analysis methods, commercial and regulatory aspects and intellectual property issues.

References

Ashley EA, Butte AJ, Wheeler MT et al (2010) Clinical assessment incorporating a personal genome. Lancet 375:1525–1535

Børresen-Dale AL (2003) Genetic profiling of breast cancer: from molecular portraits to clinical utility. Int J Biol Markers 18:54–56

Chapman P (2010) Global cancer treatment market worth $50 billion. http://www.companiesandmarkets.com/news/global-cancer-treatment-market-worth-50-billion-led-by-mab-brands-mabthera-avastin-and-herceptin-n122.aspx?prk=33f895519277bace25c442b39bd0a1e3. Accessed 1 July 2011
DeRisi JL, Iyer VR (1999) Genomics and array technology. Curr Opin Oncol 11:76–79
Glas AM, Floore A, Delahaye LJ, Witteveen AT et al (2006) Converting a breast cancer microarray signature into a high-throughput diagnostic test. BMC Genomics 7:278
Gress TM, Hoheisel JD, Lennon GG, Zehetner G, Lehrach H (1992) Hybridization fingerprinting of high-density cDNA-library arrays with cDNA pools derived from whole tissues. Mamm Genome 3:609–619
Gunderson KL, Steemers FJ, Ren H, Ng P, Zhou L, Tsan C, Chang W, Bullis D, Musmacker J, King C, Lebruska LL, Barker D, Oliphant A, Kuhn KM, Shen R (2006) Whole-genome genotyping. Methods Enzymol 410:359–376
Jain KK (2005) Applications of AmpliChip CYP45. Mol Diagn 9:119–127
Jordan BR (2004) How consistent are gene expression chip platforms? BioEssays 26:1236–1242
Jordan BR (2010) Is there a niche for DNA microarrays in molecular diagnostics? Expert Rev Mol Diagn 10:875–882
Lam CW, Lau KC, Tong SF (2010) Microarrays for personalized genomic medicine. Adv Clin Chem 52:1–18
Lennon GG, Lehrach H (1991) Hybridization analyses of arrayed cDNA libraries. Trends Genet 7:314–317
Lenoir T, Giannella E (2006) The emergence and diffusion of DNA microarray technology. J Biomed Discov Collab 1:11
Li X, Quigg RJ, Zhou J, Gu W, Nagesh Rao P, Reed EF (2008) Clinical utility of microarrays: current status, existing challenges and future outlook. Curr Genomics 9:466–474
MAQC Consortium, Shi et al (2006) The MicroArray Quality Control (MAQC) project shows inter- and intraplatform reproducibility of gene expression measurements. Nat Biotechnol 24:1151–1161
Medical News Today (2008) Global molecular diagnostics market to reach 3.67 billion dollars by 2010, according to new report by Global Industry Analysts, Inc. http://www.medicalnewstoday.com/articles/104274.php. Accessed 25 July 2011
Moser M (2001) Analysts predict red hot future for Biochips: sector poised to grow to $3.3 billion by 2004. bioArray News 1:3–4. http://www.genomeweb.com/arrays/analysts-predict-red-hot-future-biochips-sector-poised-grow-33-billion-2004. Accessed 25 July 2011
Nguyen C, Rocha D, Granjeaud S, Baldit M, Bernard K, Naquet P, Jordan BR (1995) Differential gene expression in the murine thymus assayed by quantitative hybridization of arrayed cDNA clones. Genomics 29:207–215
Schena M, Shalon D, Davis RW, Brown PO (1995) Quantitative monitoring of gene expression patterns with a complementary DNA microarray. Science 270:467–470
Simon R (2008) Lost in translation: problems and pitfalls in translating laboratory observations to clinical utility. Eur J Cancer 44:2707–2713
Tan PK, Downey TJ, Spitznagel EL Jr, Xu P, Fu D et al (2003) Evaluation of gene expression measurements from commercial microarray platforms. Nucleic Acids Res 31:5676–5684
Zhao N, Hashida H, Takahashi N, Misumi Y, Sakaki Y (1995) High-density cDNA filter analysis: a novel approach for large-scale, quantitative analysis of gene expression. Gene 156:207–213

Chapter 3
Expression Profiling for Diagnostics

Bertrand Jordan

3.1 Early Hopes and Difficulties

Determination of expression profiles has been the major research application of DNA microarrays, especially during the initial years, from 1995 on. Many studies were conducted on expression profiles in various types of cancer, with the explicit aim of differentiating between different subtypes and obtaining information of clinical importance, such as the prognosis (likelihood of recurrence or metastasis) or the prediction of efficacy for chemotherapy (Alizadeh et al. 2000; Perou et al. 2000). These studies generated great interest but also strong criticism since much of the initial work was technically flawed, performed on first-generation platforms of poor performance and reproducibility (Tan et al. 2003) or used improperly designed statistical methods for data analysis (Simon et al. 2003; Campbell 2004). Indeed, expression profiling on DNA arrays provides "noisy" data, with fairly large coefficients of variation, and there is a great imbalance between the number of genes assessed (many thousands) and the number of samples (often less than one hundred)—the so-called curse of dimensionality. Under these conditions, it is quite easy to derive classifications and expression profiles that appear to discriminate between patients and controls—only to discover, when they are tested on a new set of samples, that the classification does not hold. The profiles apparently associated with prognosis were actually based on random fluctuations; thus, they are not confirmed when a new, independent series of samples is processed. Apart from strict attention to technical details, and systematic use of replicates (often omitted in early experiments), more rigorous statistical methods must be used to avoid these pitfalls (Quackenbush 2006; Ein-Dor et al. 2006). Another difficulty is that gene expression is now known to be quite variable from individual to individual and even

B. Jordan (✉)
CoReBio PACA, Luminy Science Park, 13288 Marseille, Cedex 9, France
e-mail: bertrand.jordan@univmed.fr

B. Jordan (ed.), *Microarrays in Diagnostics and Biomarker Development*,
DOI 10.1007/978-3-642-28203-4_3, © Springer-Verlag Berlin Heidelberg 2012

at different time points in the same person, thus making interpretation of changes even more difficult. Additionally, in cancer studies, the validity of the profiles obtained can be compromised by tumour heterogeneity.

3.2 Expression Profiling Diagnostics in Actual Clinical Use

3.2.1 MammaPrint Versus Oncotype

The problems that plagued the initial studies were progressively addressed, and a number of robust profiles with proven clinical correlates were established and eventually translated into diagnostic products (see examples in Table 3.1). The best known of these is the MammaPrint test, performed by Agendia (Amsterdam, the Netherlands) and based on work initially published in 2002 (van de Vijver et al. 2002; Buyse et al. 2006). It assesses the expression of a set of 70 genes (plus many controls) to obtain prognostic information (likelihood of recurrence or metastasis) in the case of early (stage I or II) breast cancer. This test may also provide some predictive information, i.e. indicate whether or not chemotherapy is likely to be effective (Cardoso et al. 2007; Chen et al. 2010). Indeed, the distinction between *prognostic* information (how is the disease likely to develop?) and *predictive* information (is a given treatment likely to be effective for this patient?) is very important in cancer management, even though the two terms are often loosely used. In practice, Agendia performs the test using fresh-frozen tumour material and a custom microarray manufactured by Agilent and provides a classification of the patient as "low risk" or "high risk" for metastasis.

The MammaPrint test was CE IVD marked in 2004 and was the first microarray expression test to obtain FDA clearance in 2007. Although it is performed in the "home brew" format (i.e. in the company's laboratory), FDA strongly suggested formal submission because of the complexity of the analysis necessary to interpret the expression profiles determined. Indeed, proprietary algorithms and data are needed to obtain the classification of the sample, which led FDA to consider MammaPrint as an *In Vitro Diagnostic Multivariate Index Assay* (IVDMIA).[1] Adoption of the test has so far remained limited, and sales figures appear to be relatively low, around 2,000 tests in 2010,[2] although it is now covered for reimbursement in the USA for approximately 70 million individuals, i.e. 25% of the

[1] *IVDMIA* is an acronym coined by the FDA to designate array tests in which the clinical implication of the result requires the use of proprietary algorithms and cannot be derived from the data by the scientist or clinician. See Chap. 12 for a discussion of these issues.

[2] Figures based on sales information released by Agendia in documents for an IPO planned for June 2011 but subsequently withdrawn. Based on the "list price" of the test, and on the income figure quoted in the IPO documents, a large fraction of these tests must have been sold at highly discounted prices.

Table 3.1 Examples of array diagnostic products based on expression profiles

Test name	Manufacturer	Aim of test	Samples used	Technology	Approval	Cost	Adoption
Mammaprint	Agendia (The Netherlands)	Prognosis of breast cancer recurrence; selection for adjuvant chemotherapy?	Fresh-frozen tissue aliquot	70-Gene signature, Agilent array, 3,000 long oligos	CE 2005, FDA 2007	Approx. €3,000	Moderate (~2,000 tests per year); 25% insurance coverage (USA)
TOO (Tissue of origin)	Pathwork Diagnostics (USA)	Suggest primary tumour site for metastatic cancer	Fresh tissue, FFPE	2,000 Gene array	FDA 2008, FDA 2010 (FFPE)	Service €2,600, kit €1,400	New on market for FFPE
AML profiler	Skyline Diagnostics (The Netherlands)	9 fusion, mutation and expression markers to determine AML subtype	Fresh bone marrow	Custom Affymetrix GeneChip	CE 2011, FDA in progress	Various implementations €1,000–2,000	New on market
Oncotype	Genomic Health (USA)	Prognosis of breast cancer recurrence (recurrence index), prediction of chemotherapy efficacy	FFPE	Not an array, qPCR on 21 genes	510(k) (home brew)	Approx €2,500	Good (~60,000 tests per year); 95% insurance coverage (USA)

This table outlines the main features of three well-publicised expression array tests and compares them with the Oncotype PCR-based test that addresses a similar clinical issue

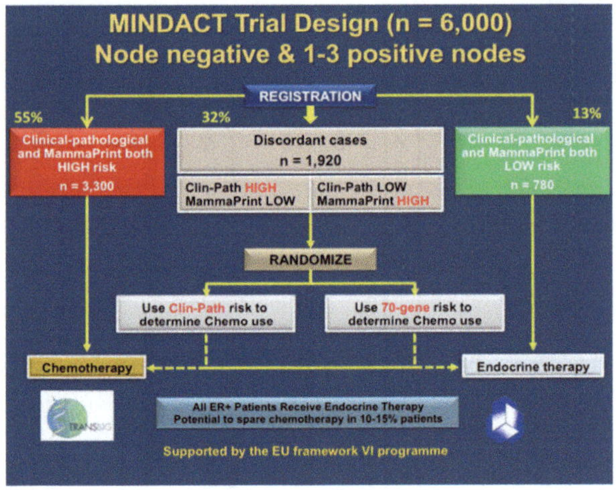

Fig. 3.1 Scheme of the MINDACT trial, designed to test the clinical utility of the MammaPrint expression profiling test performed by Agendia. Note the number of samples (very high compared to most profiling studies) and the complexity of the scheme (http://www.agendia.com/pages/mindact/128.php)

insured population. MammaPrint has probably suffered from competition by the Oncotype expression test marketed by Genomic Health (Paik et al. 2004). Also aimed at breast cancer prognosis, and providing some predictive information as well, this is not a microarray test as it assays the expression of 21 genes by real-time PCR. One of its major advantages is that the analysis can be performed on formalin-fixed paraffin-embedded ("FFPE") samples that are very widely available as they form an integral part of standard pathology practice, while MammaPrint requires fresh-frozen material, much more difficult to obtain and to handle. Oncotype has achieved wide acceptance (close to 60,000 tests sold in 2010) and is covered for reimbursement, in the USA, for almost all of the insured population. Both tests are quite expensive, in the €2,000–3,000 range.

Even though these tests are quite widely used in clinical practice, a number of authors and organisations question their usefulness (EGAPP 2009; Koscielny 2010). The issue of clinical utility (Simon 2008) goes beyond scientific validity and proven clinical correlation: the question asked here is *whether use of the test actually improves the outcome for the patient,* by allowing a more rational choice of therapy and avoiding overtreatment. Indeed, large-scale trials are under way to resolve this issue: MINDACT (Microarray in Node-Negative Disease May Avoid Chemotherapy) for MammaPrint (Cardoso et al. 2007; Mook et al. 2007) and TAILORX (Trial Assigning Individualised Options for Treatment (Rx)) for Oncotype (Zujewski and Kamin 2008). Figure 3.1 shows the scheme of the MINDACT trial and underlines its complexity. Several thousands of patients have now been enrolled in each of these clinical trials, but final results will still require several years to emerge: enrolment in MINDACT, for example, is expected to be completed in 2012, and the first results should be available late 2015 (MINDACT 2009; EORTC 2009). This extended timeline (MINDACT was initially launched in 2007) highlights the difficulty of verifying clinical utility.

3.2.2 Other Examples of Expression Profiling Diagnostic Tests

Other DNA array diagnostics based on expression profiles have also been developed and in some cases have obtained regulatory clearance. One example is the TOO (Tissue Of Origin) test (Dumur et al. 2008) developed by Pathwork Diagnostics (Redwood City, CA, USA) and approved by FDA in August 2008. This examines the expression level of 1,500 genes to determine the tissue origin of cancer that is detected at a late, metastasised stage. This information can be important for treatment, as metastases retain many of the features (some aspects of the expression profile but also of drug sensitivity) of the tissue in which the initial tumour appeared. The TOO test can be performed on FFPE samples, compares the expression profile with that of 15 reference tissues and provides a suggestion for the tissue of origin, typically with >90% probability for one particular tissue and less than 10% for the others: overall figures quoted by the company are 89% sensitivity and 99% specificity.

Additional expression tests, such as the Tumour grade proposed by Ipsogen (Marseille, France)[3] based on work in the Jules Bordet Institute (Belgium) (Sotiriou et al. 2006) or the "AML profiler" by Skyline Diagnostics (Rotterdam, the Netherlands) that ascertains alterations in acute myeloid leukaemia by expression profiles (Valk et al. 2004),[4] both CE-marked, are at the stage of market entry. A particularly interesting case is the BrochoGen test developed by Allegro Diagnostics (Maynard, MA, USA): this provides early diagnosis of lung cancer in smokers through ascertainment of an 80-gene expression profile in *airway cells,* collected by simple airway brushing (Spira et al. 2007). Contrary to the assays described above that require tumour material, this test actually detects modifications in easily accessible lung tissue that have been shown to provide a strong indication for the presence of early lung cancer. It should be marketed in 2012. One should also mention tests based on the assessment of microRNA expression, not yet used in clinical practice but being actively developed and probably quite close to medical use (Negrini et al. 2009).

3.3 Acceptance Remains Limited: A Narrow Path for Clinical Expression Arrays

On the whole, however, it has proven more difficult than expected to turn expression profiles into widely accepted clinical diagnostic tools—even though there is obviously much information in the complex patterns observed and much to be learned by interpreting them in the light of constantly evolving genomic and

[3] Currently being acquired by Qiagen.

[4] The Skyline test uses expression levels of a large number of genes to assess translocations, mutations and overexpression of critical AML markers.

functional knowledge. Apart from technical and statistical considerations, and from the relative novelty of this approach, this is largely due to the difficulty and duration of clinical studies (see above the timeline for MINDACT) that must be performed once the test is scientifically established—in contrast, again, to genotyping tests for which the clinical correlates of the mutation(s) or subtypes are generally known beforehand. In addition, the complexity of the algorithms that link expression results to clinical prognosis and prediction has motivated FDA to regulate even "home brew" tests as IVDMIAs, an additional if necessary hurdle. The recent flurry of large-scale sequencing studies on cancer samples (Wooster and Bachman 2010; Chapman et al. 2011) has revealed the rather discouraging complexity of genomic alterations present in cancer cells: thousands of point mutations as well as many insertions, deletions, copy number changes and translocations are found in the tumour cells. Expression profiles, representing as they do the actual functioning of the cancer cell, may eventually provide a better operational characterisation of the tumour's physiology and may turn out to be more useful for choice of medication than a wide catalogue of genomic alterations.

Expression arrays suffer competition mostly at the low end of the multiplicity scale. If the amount of information needed is limited (e.g. measuring the expression level of 10–20 genes), PCR-based approaches may be preferred as they are familiar to most laboratories, well understood and more easily approved by regulatory authorities and also more amenable to the use of FFPE samples, a very important point in practice, as shown by the relative success of Genomic Health's Oncotype versus Agendia's MammaPrint.

At the other end of the scale, DNA arrays are now in competition with high-throughput DNA sequencing for a number of applications that also include expression profiling. RNA-seq approaches are already superseding arrays for research expression experiments, as they generally provide much better data—at a price that is no longer prohibitive compared to arrays if many samples are incorporated (multiplexed) in a single sequencing run, after addition of suitable tags to deconvolute the results (Morozova et al. 2009). RNA-seq may eventually penetrate clinical applications, although the necessary multiplexing is a serious difficulty as it introduces unacceptable delays in many clinical situations. Arrays are well adapted to situations where short turnaround time is needed, and they are also relatively inexpensive, at least compared to current sequencing costs. Miniaturisation ("nanoarrays"), automation of the assays and the use of unlabelled samples are all avenues that could increase their success in the diagnostic field (see Chap. 10).

References

Alizadeh AA, Eisen MB, Davis RE, Ma C, Lossos IS et al (2000) Distinct types of diffuse large B-cell lymphoma identified by gene expression profiling. Nature 403:503–511

Buyse M, Loi S, van't Veer L, Viale G, Delorenzi M et al (2006) Validation and clinical utility of a 70-gene prognostic signature for women with node-negative breast cancer. J Natl Cancer Inst 98:1183–1192

Campbell G (2004) Some issues in the statistical evaluation of genetic and genomic tests. J Biopharm Stat 14:539–552

Cardoso F, Piccart-Gebhart M, Van't Veer L, Rutgers E, TRANSBIG Consortium (2007) The MINDACT trial: the first prospective clinical validation of a genomic tool. Mol Oncol 1:246–251

Chapman MA, Lawrence MS, Keats JJ, Cibulskis K, Sougnez C et al (2011) Initial genome sequencing and analysis of multiple myeloma. Nature 471:467–472

Chen E, Tong KB, Malin JL (2010) Cost-effectiveness of 70-gene MammaPrint signature in node-negative breast cancer. Am J Manag Care 16:e333–42

Dumur CI, Lyons-Weiler M, Sciulli C, Garrett CT, Schrijver I et al (2008) Interlaboratory performance of a microarray-based gene expression test to determine tissue of origin in poorly differentiated and undifferentiated cancers. J Mol Diagn 10:67–77

EGAPP (Evaluation of Genomic Applications in Practice and Prevention Working Group) (2009) Recommendations from the EGAPP Working Group: can tumor gene expression profiling improve outcomes in patients with breast cancer? Genet Med 11:66–73

Ein-Dor L, Zuk O, Domany E (2006) Thousands of samples are needed to generate a robust gene list for predicting outcome in cancer. Proc Natl Acad Sci USA 103:5923–5928

EORTC (2009) http://www.eortc.be/services/unit/mindact/MINDACT_websiteii.asp#Current. Accessed 28 July 2011

Koscielny S (2010) Why most gene expression signatures of tumors have not been useful in the clinic. Sci Transl Med 2:14 ps2

MINDACT progress report (2009) http://www.breastinternationalgroup.org/LinkClick.aspx?fileticket=UsF4HCJsbKQ%3d&tabid=2193. Accessed 25 July 2011

Mook S, Van't Veer LJ, Rutgers EJ, Piccart-Gebhart MJ, Cardoso F (2007) Individualization of therapy using Mammaprint: from development to the MINDACT Trial. Cancer Genomics Proteomics 4:147–155

Morozova O, Hirst M, Marra MA (2009) Applications of new sequencing technologies for transcriptome analysis. Annu Rev Genomics Hum Genet 10:135–151

Negrini M, Nicoloso MS, Calin GA (2009) MicroRNAs and cancer–new paradigms in molecular oncology. Curr Opin Cell Biol 21:470–479

Paik S, Shak S, Tang G, Kim C, Baker J et al (2004) A multigene assay to predict recurrence of tamoxifen-treated, node-negative breast cancer. N Engl J Med 351:2817–2826

Perou CM, Sørlie T, Eisen MB, van de Rijn M, Jeffrey SS (2000) Molecular portraits of human breast tumours. Nature 406:747–752

Quackenbush J (2006) Microarray analysis and tumor classification. N Engl J Med 354:2463–2472

Simon R (2008) Lost in translation: problems and pitfalls in translating laboratory observations to clinical utility. Eur J Cancer 44:2707–2713

Simon R, Radmacher MD, Dobbin K, McShane LM (2003) Pitfalls in the use of DNA microarray data for diagnostic and prognostic classification. J Natl Cancer Inst 95:14–18

Sotiriou C, Wirapati P, Loi S, Harris A, Fox S et al (2006) Gene expression profiling in breast cancer: understanding the molecular basis of histologic grade to improve prognosis. J Natl Cancer Inst 98:262–272

Spira A, Beane JE, Shah V, Steiling K, Liu G et al (2007) Airway epithelial gene expression in the diagnostic evaluation of smokers with suspect lung cancer. Nat Med 13:361–366

Tan PK, Downey TJ, Spitznagel EL Jr, Xu P, Fu D et al (2003) Evaluation of gene expression measurements from commercial microarray platforms. Nucleic Acids Res 31:5676–5684

Valk PJ, Verhaak RG, Beijen MA, Erpelinck CA, van Waalwijk B, van Doorn-Khosrovani S et al (2004) Prognostically useful gene-expression profiles in acute myeloid leukemia. N Engl J Med 350:1617–1628

van de Vijver MJ, He YD, van't Veer LJ, Dai H, Hart AA et al (2002) A gene-expression signature as a predictor of survival in breast cancer. N Engl J Med 347:1999–2009

Wooster R, Bachman KE (2010) Catalogue, cause, complexity and cure; the many uses of cancer genome sequence. Curr Opin Genet Dev 20:336–341

Zujewski JA, Kamin L (2008) Trial assessing individualized options for treatment for breast cancer: the TAILORx trial. Future Oncol 4:603–610

Chapter 4
Genotype Characterisation for Diagnostics

Bertrand Jordan

4.1 Simple Genotyping Arrays

"Simplicity" is understood here as the number of locations or spots ("features") on the array—simple devices in this sense may contain up to one or two hundred spots (including duplicates and controls). With such arrays, it is possible to perform, in a single step, fairly sophisticated characterisation of a pathogen or to assay simultaneously for several mutations in a particular gene—thus providing significantly more information than a PCR-based assay while retaining relatively simple apparatus and accessible cost, typically of the order of €100 per sample. For example, detailed typing of HPV (human papillomavirus) found in a cervical smear sample has important clinical significance, as some HPV subtypes are associated with a high risk of cervical cancer, while others are largely innocuous (Leinonen et al. 2009). Currently, HPV typing is becoming incorporated in routine clinical practice and may in the future replace, or systematically complement, the cytological examination (cervical smear, or "Pap test") that is widely used at present, with over 50 million tests per year in the USA alone (National Cancer Institute 2010). More than a hundred HPV subtypes have been described, but assessing for the most common 10–15 is normally sufficient. This makes such ascertainment quite a good prospect for a simple array, and indeed a number of manufacturers are competing in this market. The PapilloCheck system (see Table 4.1) developed by Greiner Bio-One (Frickenhausen, Germany) is a good example. It is based on a 140-feature DNA microarray that contains, in addition to a number of control spots, oligonucleotides specific for the 24 major papillomavirus subtypes (spotted five times to improve precision). Material corresponding to a fragment of the E1 gene in HPV is amplified from a cervical smear sample by PCR and labelled with the Cy5 fluorophore, and then hybridised (together with control, Cy3-labelled material) to

B. Jordan (✉)
CoReBio PACA, Luminy Science Park, 13288 Marseille, Cedex 9, France
e-mail: bertrand.jordan@univmed.fr

B. Jordan (ed.), *Microarrays in Diagnostics and Biomarker Development*,
DOI 10.1007/978-3-642-28203-4_4, © Springer-Verlag Berlin Heidelberg 2012

Table 4.1 Examples of "simple" (low-multiplex) genotyping diagnostic products

Test name	Manufacturer	Aim of test	Samples used	Technology	Approval	Cost	Adoption
PapilloCheck	Greiner Bio-One (Germany)	Determination of HPV subtype in cervical smear samples	Cervical smear	140-Oligonucleotide array (multiple spotting)	CE-IVD 2009	Below €20 per sample	Fairly good (strong competition)
eSensor cystic fibrosis genotyping test	GenMark Dx (USA), formerly Osmetech (UK)	Assay the 23 most common CF mutations	Blood	Oligonucleotide array, eSensor technology (no sample labelling)	FDA 2006, 2008	Below €100 per sample	Moderate (very strong competition)
BloodChip	Progenika Biopharma, Spain	Identification of 33 blood groups and 12 platelet antigens	Blood	Sendagene platform (proprietary oligonucleotide chip)	CE 2010	Approx. €175 per sample	Entering market

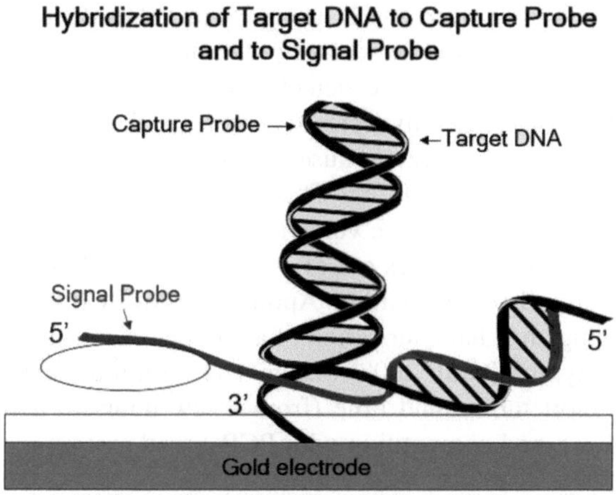

Fig. 4.1 The GenMark/Osmetech label-free detection system. The patient's DNA is first hybridised to the array, then the "signal probe" hybridises to the overhang of the (PCR-amplified) target DNA bound to the "capture probe" and is detected electrically, thanks to its ferrocene label (oval at 5′ end). Figure adapted from Osmetech promotional material

the slide. Results are acquired by scanning the slide after washing. Total assay time, including PCR amplification, is approximately 2 h, and the results are qualitative (presence or absence of a given HPV subtype), not quantitative (abundance of each subtype in mixed samples) because of the PCR step. This test is sold as 12-sample slides, and the cost per sample is below €20, quite a performance. It has obtained CE-IVD status but is not FDA-approved at this time.

Another type of simple array is directed at identifying a limited set of possible mutations in a particular human gene. A major case here is the *CFTR* gene responsible (when mutated) for cystic fibrosis, a disease for which the carrier frequency approaches 1/20 in Western populations and that represents approximately 20% of the microarray diagnostics market. For this disease, in addition to the major deltaF501 mutation, a number of other mutations present in the population at significant levels have been described (Moskowitz et al. 2008). A useful screening test should therefore be able to detect a set of possible mutations; clinical advisory bodies have defined the 23 most frequent of these, to be assessed in first-round screening (Watson et al. 2004), and this choice has been confirmed recently (Schrijver 2011). Several companies have developed tests aimed at these mutations (Johnson et al. 2007), sometimes with technologies specially adapted to a clinical environment. For example, Osmetech [UK, now GenMark Diagnostics (Pasadena, CA, USA)] has implemented a technology originally invented at Motorola (Austin, USA) in which the detection relies on an electrical signal (without sample labelling) and accordingly allows for a high degree of automation of the procedure (GenMark Dx eSensor® Cystic Fibrosis Genotyping Test 2011). Essentially (Fig. 4.1), the sample is hybridised with a "capture probe" placed in a microfluidic cartridge that includes a sequence complementary to part of the expected sample sequence (after PCR amplification). Then a "signal probe" complementary to the free sample sequence and carrying a ferrocene label is introduced into the cartridge. The ferrocene label is detected by the response to voltage increase across the electrode. Up to 72 analytes can be detected with a single cartridge. This is a good example of the user-friendly detection systems that can be developed for genotyping DNA arrays, largely because the sensitivity requirements are less stringent than for expression profiling. This test has obtained FDA clearance in 2006 and 2009 for its two successive versions (see Table 4.1).

The above are just two examples out of a large number of "simple" genotyping array-based tests currently on the market. Applications of such systems also include influenza virus typing and characterisation of a number of infectious agents as well as HLA and blood typing (Quill 2008) (Table 4.1). Common features of these are low multiplicity, short turnaround time (from a few hours to a couple of days), relatively low cost . . . and competition with PCR-based assays.

4.2 Complex Genotyping Arrays

4.2.1 Detecting Many Mutations

Complex arrays contain thousands (or more) of oligonucleotide probes and aim at identification of complex pathogens (or mixtures thereof) or, alternatively, at detailed investigation of mutations for several genes. They provide a large amount of information but require sophisticated instrumentation and carry accordingly higher cost (typically at least 500 USD). The prime example here is the Roche/Affymetrix AmpliChip CYP450 test, which was the first "real" microarray test to obtain FDA clearance in 2004 and 2005 (Jain 2005). It uses a specific 15,000 oligonucleotide chip manufactured by Affymetrix and assesses a number of mutations in the *CYP2D6* and *CYP2C19* genes that are involved in the metabolism of many psychoactive drugs and are known to have "slow metabolising" and "fast-metabolising" variants in the human population (Table 4.2). A slow metaboliser will degrade the drug slowly and thus experience very strong effects at the regular dose, while a fast metaboliser will require a high dose to achieve the standard effect (Daly 2003). Performing the test *before* treatment should allow individual adjustment of medication and avoid over- and under-treatment for those rare patients (a few per cent) who do not carry the "standard" version of the *CYP2D6* and *CYP2C19* genes. Although sales figures for this test are not available, it seems to have been only moderately successful, both because of its high cost (approx. €1,000 initially per assay) and of the fact that its predictive value has been questioned (Katsanis et al. 2008).

In a different implementation, oligonucleotides corresponding to a number of "simple" arrays can be combined into a more complex device in order to allow highly multiplexed diagnostics. A potentially important application is systematic screening for carrier status for a large number of monogenic diseases, offered to

Table 4.2 Examples of complex genotyping diagnostic products

Test name	Manufacturer	Aim of test	Samples used	Technology	Approval	Cost	Adoption
AmpliChip CYP450	Roche Molecular Diagnostics	Gene variations—including deletions and duplications—for the *CYP2D6* and *CYP2C19* genes	Blood	15,000 Oligonucleotide Affymetrix chip	FDA 2004, 2005	600–1,300 USD	Moderate
Universal carrier test	Counsyl (USA)	Carrier screen for 100+ single gene disorders	Saliva	Oligonucleotide array, 925 probes for 458 variants	LDT (CLIA laboratory)	349 USD	On market end 2009
Human Genomic CGH Microarray ISCA V2	Agilent Technologies (USA)	CNV detection at medium resolution (whole genome)	Blood	44,000 60-mer oligos (4 arrays per slide)	RUO but incorporated in clinical evaluations	<100 USD per sample	Good (various versions)

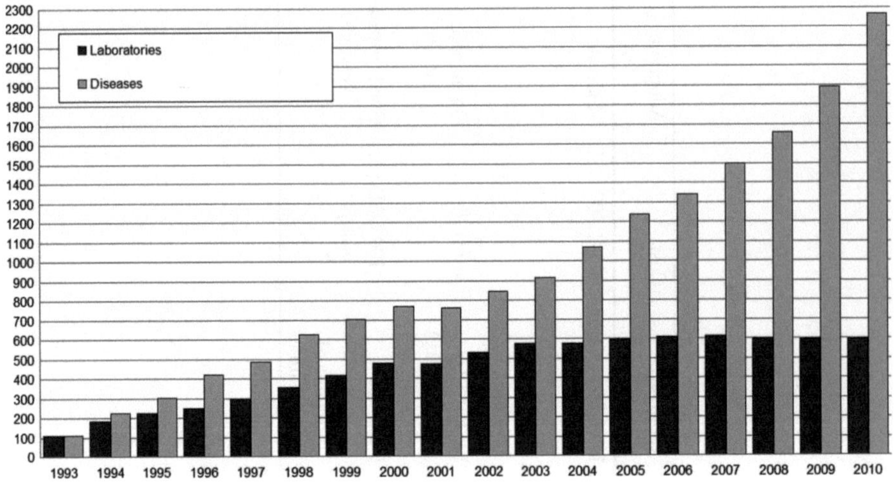

Fig. 4.2 Growth in the number of gene tests available in the USA. *Grey bars*: number of gene tests available; *black bars*: number of registered laboratories. Adapted from the GeneTests website

couples before reproduction. The aim here is to consolidate the very large number of existing tests (~2,000) assaying for mutations in particular genes (Fig. 4.2) (GeneTests 2010; Bell et al. 2011) into a single one assessing for all known recessive monogenic diseases. The expectation is that if both prospective parents are carriers for a given disease, they will resort to in vitro fertilisation and preimplantation genetic diagnosis in order to conceive a healthy child—an expensive but quite workable solution, at least in high-income countries. A somewhat similar scheme has been implemented within the Jewish Ashkenazi community (with carefully thought out ethical safeguards) and has essentially eliminated the Tay-Sachs disease previously quite common in that community because of founder effects (Kaback 2000). At least one company is providing such an array-based test, screening the parent's DNA for the most common mutations in more than 100 genetic diseases (Counsyl 2011; Srinivasan et al. 2010) at a relatively accessible cost (Table 4.2), using DNA array technology. However, as the number of genes and mutations to be assessed increases, sequencing using NGS becomes an increasingly attractive alternative, and this particular application may well in the future be taken over by sequencing (Bell et al. 2011). Other complex tests examine, for example, a number of genes involved in cardiomyopathies using custom Affymetrix arrays (Partners Healthcare 2011), or provide detailed information on blood subgroups and HLA alleles in the context of transfusion and transplantation therapy (Progenika Biopharma 2011). There as well, some of these tests are being moved over to sequencing technology.

4.2.2 Scoring Copy Number Variation

Another type of complex genotyping array has recently been assuming a growing role in clinical studies. Variously called "CGH array", "CytoChip" or "CNV array", this is designed to assay for the presence of copy number variation (hence CNV) in a patient's DNA (Veltman 2006; Bejjani and Shaffer 2006). Such arrays contain thousands of probes distributed along the human genome and are hybridised with a mixture of normal human DNA (labelled with one fluorophore) and a patient's DNA (labelled with a different fluorophore). When anomalies are present in the patient's DNA, the two signals differ strongly in intensity and thus reveal deletions or duplications (more generally, CNVs) in the patient's DNA under study.[1] This approach replaces an earlier technique, called "comparative genomic hybridisation" that used a similar DNA mix for hybridisation to metaphase chromosomes and thus revealed copy number changes—but of course with much cruder resolution. Given the frequent occurrence of CNVs in human DNA, and the fact that many of them have been linked to various diseases, "CytoChips" are on the way to complement or even replace conventional cytogenetic examinations (karyotypes) that are very widely practised but are both labour intensive and of low resolution. Since they probably constitute at this time the most important clinical application of DNA arrays, they will be discussed in detail in the following chapter (Chap. 5).

4.2.3 General Genotyping Arrays in Research and "Recreational Genetics"

In terms of research applications, scoring single nucleotide polymorphisms (SNPs or snips) is probably now the major application of DNA arrays: current offerings from several manufacturers (Illumina, Affymetrix, Agilent) allow scoring for 1,000,000–2,000,000 snips in a patient's DNA in a single experiment and have found very wide use in genome-wide association studies (GWAS) (McCarthy et al. 2008). These studies have indeed allowed reliable identification of a number of loci and genes that show linkage to vulnerability in complex diseases, leading to better understanding of the aetiology of these conditions and possibly to new therapeutic approaches. However, the loci thus identified, when taken together, account for only a small fraction of the known heritability of the conditions and have therefore little predictive value (Manolio et al. 2009). Accordingly, such studies—while very valuable in research—have little direct clinical relevance, yet a number of companies offer such services in "direct-to-consumer" (DTC) format. These firms

[1] Array CGH was initially implemented using arrays of DNA derived from BACs (bacterial artificial chromosomes). This approach is now largely phased out due to low resolution and to manufacturing issues.

perform genome-wide genotyping on their customers' DNA, often using 500,000 snip chips, and provide the results together with an interpretation in terms of susceptibility to a set of twenty or more common diseases. Although, to avoid regulation, these companies insist that they are not providing medical information *stricto sensu*, the implication is that customers will use the data to adjust their lifestyle (and in some cases undergo specific exploration procedures) in order to minimise their risk. The actual value of this information, based on the "common variant/common disease" hypothesis (Reich and Lander 2001), is however very limited, as shown in particular by a recent comparative study (Government Accountability Office 2010). The situation may progressively change with increasing precision of genomic information (sequencing) and better knowledge of genotype to phenotype correlations, but the change will be slow and, at this time, these exercises in "recreational genetics" cannot be considered to belong in the diagnostic field.

In summary, complex genotyping arrays have indeed found significant applications in clinical practice. As emphasised previously, they are generally building on previously established correlations between DNA sequences and disease and usually do not require further investigation on clinical validity and utility, although the CYP450 case shows that this can still be an issue. As discussed below, they are likely to face strong competition from high-throughput sequencing in the near future, especially since their cost is relatively high, often in the €1,000–3,000 range.

4.3 A Narrow Path for Genotyping Arrays

Genotyping DNA arrays enjoy a relatively smooth path to regulatory approval, as the clinical correlates of the information they provide (pathogen characterisation or mutation in patient genes) are generally clear and widely accepted. On the other hand and as in the case of expression profiling, they suffer from competition at both ends of the complexity scale—indeed the competition is even more severe than for expression arrays. If the amount of information needed is limited (e.g. checking for a few mutations), PCR-based approaches are usually preferred as they are familiar to most laboratories, well understood and more easily approved by regulatory authorities. A DNA array can provide considerably more information at a cost that is only slightly higher—but clinicians may not need, or even want, this additional information. For example, CGH arrays (see Chap. 5) may detect in a patient's DNA many CNVs (some of them novel) whose clinical significance is unknown and that will therefore not be reported to the patient or even to the physician. Price-wise, it is difficult to market a (simple) array-based test at less than 30–40 USD per test, while the reagent kit for a PCR test may cost as little as 5–10 USD, a significant difference in clinical operations.

At the other end of the scale, DNA arrays are now in competition with high-throughput DNA sequencing for applications that require high multiplicity. Tremendous progress has been made in DNA sequencing, with costs that are currently below 5,000 USD for a full human genome sequence of reasonable quality (Metzker 2010). Systematic carrier screening is a good example: while screening for 100 monogenic diseases is currently offered for less than 400 USD using arrays (Counsyl 2011; Srinivasan et al. 2010), projects are under way that aim to make available sequencing-based screening for 448 diseases for a similar amount (Bell et al. 2011), and in fact, one company, Ambry Genetics (Aliso Viejo, USA), already provides sequencing-based carrier testing for 90 diseases at a similar price (Ambry Genetics 2010). Inexpensive sequence data, however, typically require multiplexing of the sequencing run, i.e. simultaneously running 50–100 suitably tagged samples in the system, an operating mode that introduces unacceptable delays in many clinical situations.

Thus, the niche for genotyping arrays corresponds to a limited set of applications, typically "mid-plex" determinations assessing less than 100 mutations. Arrays are adapted to situations where short turnaround time is needed, and they are also relatively inexpensive, at least compared to current sequencing costs. Here as well as for expression arrays, miniaturisation, automation of the assays and the use of unlabelled samples are avenues that could increase their success in the diagnostic field (see Chap. 10).

References

Ambry genetics (2010) http://www.ambrygen.com/AmbryScreen-News.html. Accessed 26 July 2011

Bejjani BA, Shaffer LG (2006) Application of array-based comparative genomic hybridization to clinical diagnostics. J Mol Diagn 8:528–553

Bell CJ, Dinwiddie DL, Miller NA, Hateley SL, Ganusova EE et al (2011) Carrier testing for severe childhood recessive diseases by next-generation sequencing. Sci Transl Med. 3:65ra4

Counsyl (2011) http://www.counsyl.com. Accessed 26 July 2011

Daly AK (2003) Pharmacogenetics of the major polymorphic metabolizing enzymes. Fundam Clin Pharmacol 17:27–41

GeneTest (2010) http://www.ncbi.nlm.nih.gov/sites/GeneTests/. Accessed 30 July 2011

GenMarkDx eSensor® Cystic Fibrosis Genotyping Test (2011) http://www.genmarkdx.com/products/reagents/cf.php. Accessed 26 July 2011

Government Accountability Office (2010) Direct-to-consumer genetic tests: misleading test results are further complicated by deceptive marketing and other questionable practices. GAO-10-847TT. http://www.gao.gov/new.items/d10847t.pdf. Accessed 30 July 2011

Jain KK (2005) Applications of AmpliChip CYP450. Mol Diagn 9:119–127

Johnson MA, Yoshitomi MJ, Richards CS (2007) A comparative study of five technologically diverse CFTR testing platforms. J Mol Diagn 9:401–407

Kaback MM (2000) Population-based genetic screening for reproductive counseling: the Tay-Sachs disease model. Eur J Pediatr 159:S192–S195

Katsanis SH, Javitt G, Hudso K (2008) A case study of personalized medicine. Science 320:53–54

Leinonen M, Nieminen P, Kotaniemi-Talonen L et al (2009) Age-specific evaluation of primary human papillomavirus screening vs conventional cytology in a randomized setting. J Natl Cancer Inst 101:1612–1623

Manolio TA, Collins FS, Cox NJ, Goldstein DB, Hindorff LA et al (2009) Finding the missing heritability of complex diseases. Nature 461:747–753

McCarthy MI, Abecasis GR, Cardon LR, Goldstein DB, Little J et al (2008) Genome-wide association studies for complex traits: consensus, uncertainty and challenges. Nat Rev Genet 9:356–369
Metzker ML (2010) Sequencing technologies – the next generation. Nat Rev Genet 11:31–46
Moskowitz SM, Chmiel JF, Sternen DL, Cheng E, Gibson RL et al (2008) Clinical practice and genetic counseling for cystic fibrosis and CFTR-related disorders. Genet Med 10:851–868
National Cancer Institute, Pap test (2010) http://www.cancer.gov/cancertopics/factsheet/Detection/Pap-test. Accessed 26 July 2011
Partners Healthcare (2011) http://www.partners.org/. Accessed 31 July 2011
Progenika Biopharma (2011) http://www.progenika.com/eu/. Accessed 31 July 2011
Quill E (2008) Blood-matching goes genetic. Science 319:1478–1479
Reich DE, Lander ES (2001) On the allelic spectrum of human disease. Trends Genet 17:502–510
Schrijver I (2011) Mutation distribution in expanded screening for cystic fibrosis: making up the balance in a context of ethnic diversity. Clin Chem 57:799–801
Srinivasan BS, Evans EA, Flannick J, Patterson AS, Chang CC et al (2010) A universal carrier test for the long tail of Mendelian disease. Reprod Biomed Online 21:537–551
Veltman JA (2006) Genomic microarrays in clinical diagnosis. Curr Opin Pediatr 18:598–603
Watson MS, Cutting GR, Desnick RJ, Driscoll DA, Klinger K et al (2004) Cystic fibrosis population carrier screening: 2004 revision of American College of Medical Genetics mutation panel. Genet Med 6:387–391

Chapter 5
Array-CGH and SNP-Arrays, the New Karyotype

Karoly Szuhai

5.1 Introduction

Array-comparative genomic hybridisation (array-CGH) and single nucleotide polymorphism array (SNP-array)-based whole-genome screening techniques have become essential tools in cytogenetic research and diagnostics. Both approaches provide a high-resolution screening for genome-wide copy number alterations (CNAs). The discovery of previously cryptic alterations associated with constitutional or acquired cytogenetic changes quickly led to the identification of novel syndromes or previously unseen tumour-related alterations and to the discovery of genomic copy number variations. The platforms are using either immobilised large genomic insert hybridised against differentially labelled test and reference samples or immobilised oligonucleotide sampling for single nucleotide polymorphism (SNP) in the genome with second read-out for the actual copy number of the investigated genomic loci in a single detection colour. Both approaches have been quickly adapted and applied to diagnostic settings and were found to be superior to conventional chromosome-based cytogenetic investigation.

5.2 Outline of Cytogenetic Approaches

5.2.1 Conventional Cytogenetics

Traditionally, cytogenetic analysis of G-banded metaphase chromosomes was used to detect numerical and structural chromosomal aberrations. Genomic changes involving chromosomal aneuploidies (gain or a loss of a whole chromosome) and structural

K. Szuhai (✉)
Molecular Cell Biology, Leiden University Medical Center, Leiden, the Netherlands
e-mail: k.szuhai@lumc.nl

B. Jordan (ed.), *Microarrays in Diagnostics and Biomarker Development*,
DOI 10.1007/978-3-642-28203-4_5,

alterations (deletion, duplications, translocations and marker chromosomes) are an underlying cause of both constitutional (bodily dysmorphism, congenital anomalies, miscarriages and several other genetic syndromes) and acquired (benign or malignant tumour formation, secondary changes as a result of environmental exposure such as ionising radiation exposure) anomalies. However, the resolution of a routine karyotype is limited to 5–10 Mb depending on (a) chromosomal location of the breakpoints (the detection based on changes in the length and or banding pattern differences), (b) the quality of the chromosome preparation and (c) the skill and experience of cytogeneticist.

5.2.2 Targeted DNA-Based Approaches

With the introduction of fluorescence in situ hybridisation (FISH), using labelled DNA fragments as probes to interrogate different genomic loci, applied to either metaphase chromosomes or interphase nuclei, the detection of submicroscopic (cryptic) chromosomal changes became possible. Until recently, FISH had been considered the method of choice for detecting chromosomal imbalances and rearrangements in both constitutional and acquired conditions. FISH is routinely used to identify and verify chromosomal alterations, to confirm the clinical suspicion of known microdeletion and duplication syndromes or to detect tumour-associated secondary changes that might be relevant for the prognosis or the selection of treatment modalities. FISH analysis is, however, a targeted method that requires prior knowledge of the genomic region(s) and therefore can only perform a screen for individual DNA targets rather than a genome-wide screen for unexpected alterations. To overcome some of these limitations, multicolour FISH-based karyotyping (SKY, MFISH, COBRA FISH) and multiplexed locus-specific probe detection methods were developed. Other molecular methods, such as multiplex amplifiable probe hybridisation (MAPH), multiplex ligation-dependent probe amplification (MLPA) and quantitative PCR, have been developed to measure the copy number status of genomic loci with the possibility to detect multiple targets in a single experiment (Emanuel and Saitta 2007; Speicher and Carter 2005).

5.2.3 Comparative Genomic Hybridisation

Another approach to perform genome-wide screen detection of copy number alteration was developed in 1992, initially for tumour cytogenetics, and dubbed metaphase comparative genomic hybridisation (CGH) (Kallioniemi et al. 1992). In CGH, differentially labelled test and reference genomic DNA samples are co-hybridised onto a normal metaphase. The resolution of this technique, basically, is determined by the quality of metaphase spreads used for the test and remains in the ~3–10-Mb range. CGH was proven to be a very useful research tool to study cancer genome-related alterations, but the low resolution of the metaphase chromosomes and technical challenges limited the widespread application of this methodology (Kirchhoff et al. 2001; Speicher and Carter 2005).

5.2.4 Array-Based Comparative Genomic Hybridisation/Single Nucleotide Polymorphism Array

5.2.4.1 Array-CGH

Thanks to the replacement of metaphase chromosomes with sequenced and mapped target DNA sequences derived from large genomic insert containing clones, or oligonucleotides, both arrayed and immobilised onto glass slides, the limitations related to resolution could be overcome and the analysis procedure could be simplified. The technology was first introduced as "matrix-CGH" and later became widely used as "array-CGH" (Pinkel et al. 1998; Solinas-Toldo et al. 1997) using large genomic insert clones as reporters. Later, microarrays containing short oligonucleotides (typically 25–60 mers) as reporter probes have been introduced by various vendors. Thanks to technological advances, the number of reporter elements on array chips increased from several hundred to over two million target oligonucleotides. To identify copy number gain or loss (Fig. 5.1), test and reference of whole genomic DNA are differentially labelled with fluorophore-conjugated oligonucleotides using either an enzymatic or a direct chemical reaction (Pinkel et al. 1998; Raap et al. 2004). Labelled DNA samples are then mixed and co-hybridised on the microarray. Following hybridisation, the slide is scanned to establish fluorescence intensities at each target on the array. Normalised, corrected fluorescence ratio values are then ordered based on their genomic position and displayed (Fig. 5.1).

The array-CGH approach, similar to classical CGH, has initially been used to analyse copy number changes in tumour with the aim of identifying genes involved in the pathogenesis of cancers (Albertson and Pinkel 2003).

5.2.4.2 SNP-Array

The SNP-array consists of an array of short oligonucleotide probes (20–30 mer) that uses mismatch hybridisation or single-base extension to detect SNP. These are single-channel arrays, meaning that only test DNA samples are labelled and hybridised. Consequently, an SNP-array can interrogate both the copy number and heterozygosity state of genomic loci, allowing allelotyping. Not only the detection of loss of heterozygosity (LOH) associated with genomic loss but also the ascertainment of copy-neutral anomalies such as uniparental disomy or homozygosity becomes

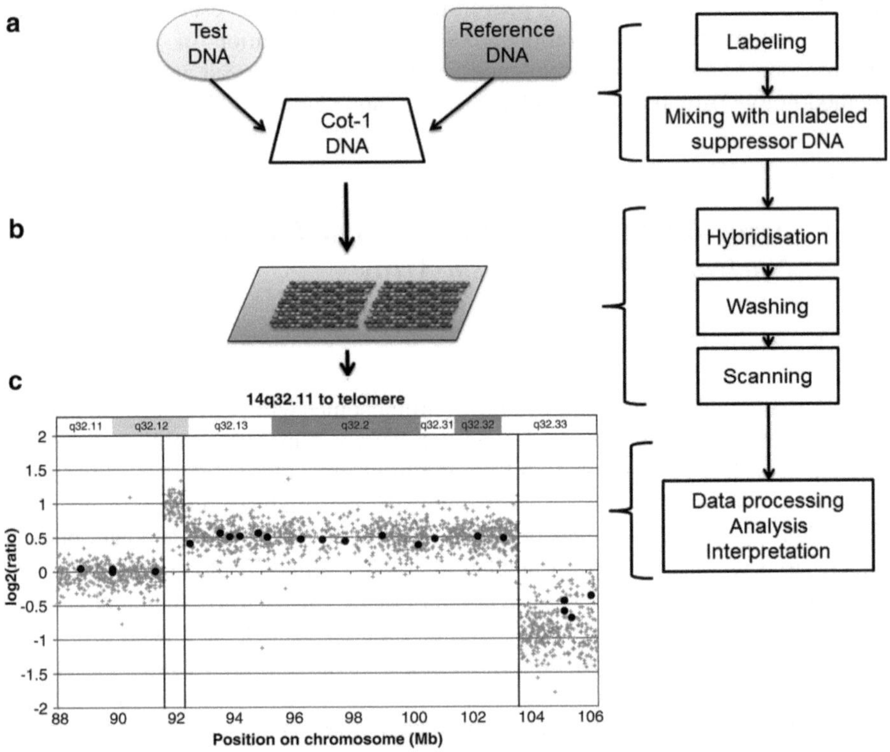

Fig. 5.1 Overview of array-based testing procedure. (**a**) The differentially labelled test and reference DNA samples are mixed with non-labelled suppressor DNA enriched for repetitive DNA sequences. (**b**) Precipitated, resuspended DNA mix is hybridised to an array chip. After hybridisation, by using stringent washing steps non-bound and partially bound probes are removed. Array chips are scanned by a laser scanner. (**c**) Fluorescence intensity values are determined to each spot, and normalisation and segmentation steps are done using various software applications. Finally, ratio values of the individual reporters are displayed according to their genomic position

feasible with SNP-arrays. The coverage of SNP probes in CNV regions is poorer compared to conventional oligonucleotide-based array-CGH approaches. To overcome this limitation of the SNP-array platforms, manufacturers have introduced probe set specifically designed for the detection of CNV. These probes are non-polymorphic reporters specially designed for the detection of CNV, resulting in more even probe coverage for the detection of copy number alterations. For example, approximately half of the reporter elements on the human Affymetrix SNP 6.0 array with 1.8 million markers are dedicated to CNV detection. For the interpretation of SNP-array, a panel of reference normal sample profile needs to be collected. To reach a firm conclusion on the SNP heterozygosity state of acquired genomic alterations, the use of both tumour- and normal tissue-derived DNA is necessary.

5.3 Applications in Clinical Practice

5.3.1 Molecular Karyotyping

It soon became evident that even a lower resolution array-CGH approach, using about 3,500 large genomic insert clones, has far better resolution than conventional G-banding of metaphase chromosomes. The high-resolution, genome-wide screening approach, either using the array-CGH or SNP-array technology, has been dubbed molecular karyotyping (Vermeesch et al. 2005).

Previously undetected, submicroscopic unbalanced constitutional rearrangements have been detected in several studies. The diagnostic yield in postnatal diagnosis of developmental delay (DD), mental retardation (MR) and/or multiple congenital anomalies (MCA) using classical metaphase analysis and G-banded chromosomes is approximately 3.7%, and an additional 2.5–7% increase has been achieved by FISH analysis using subtelomeric probe sets (Flint et al. 1995; Flint and Knight 2003). Initial studies using array-CGH with genome-wide coverage with DD/MR and MCR with a normal karyotype resulted in an additional ~15% diagnostic yield by detecting de novo copy number alterations (CNA) (Shaw-Smith et al. 2004; Vissers et al. 2003). Follow-up studies have found de novo alterations in 10–16% of cases (Stankiewicz and Beaudet 2007).

From these studies, it has become clear that the detection rate of pathogenic imbalances slightly increases by using larger number of reporter elements. About 20% of the detected genomic imbalances were smaller than 1 Mb in size, indicating that increasing the number of reporter element should result in a higher diagnostic yield (Stankiewicz and Beaudet 2007). On the other hand, the increased resolution uncovered numerous variation of DNA copy number (gains and losses) scattered all over the human genome (copy number variation: CNV) (Freeman et al. 2006; Redon et al. 2006b; Shinawi and Cheung 2008). While most of the genomic copy number differences are apparently benign CNVs (Eichler et al. 2007; Lee et al. 2007), a careful interpretation is necessary to identify potential pathogenic associations (Cooper et al. 2011; Knijnenburg et al. 2009; Sebat et al. 2007).

5.3.2 Databases/Catalogues of Genomic Alterations

In the clinical setting, it is very important to use carefully catalogued, publicly available databases listing CNVs to assist in deciding about the clinical significance of the detected genomic alterations (Database of Genomic Variants DGV http://www.projects.tcag.ca/variation/).

The overall recurrence rate of the detected deletion and/or duplication is very low in these studies, indicating that there is a great degree of genomic heterogeneity of MR and DD with or without MCA (Edelmann and Hirschhorn 2009; Lee et al. 2007). The wide variation and low frequency of recurrent alterations often make the genotype–phenotype correlation uncertain and therefore necessitate the establishment of transnational databases, collectively reporting genomic alteration from several institutions. In this way, identification of patients with common genomic rearrangements and overlapping phenotypes facilitates rapid characterisation of novel genomic syndromes. Such examples are the Database of Chromosomal Imbalance and Phenotype in Humans using Ensembl Resources (DECIPHER Consortium) (http://decipher.sanger.ac.uk/) and the European Cytogeneticists Association Register of Unbalanced Chromosome Aberrations (ECARUCA) (http://www.ecaruca.net).

5.3.3 Novel Syndromes

Before molecular cytogenetic approaches, novel syndromes were mainly identified by collecting patient cohorts with a common genomic alteration and similar phenotypic features. In many instances, reporting an index case could lead to the identification of novel syndromes by pointing to the commonly involved chromosomal region or gene. Molecular karyotyping allowing the simultaneous interrogation of several tens of thousands of discrete genomic locations for their copy number status has been successfully applied in the identification of novel syndromes. A list of novel syndromes that have been discovered by molecular karyotyping is shown in Table 5.1. The use of high-resolution molecular karyotyping allowed the identification of several, novel microduplication syndromes reciprocal to well-known microdeletion syndromes. Segmental duplication can be hot spots for the occurrence of non-allelic homologous recombination or unequal crossing-over leading to genomic mutations such as deletion, duplication, inversion or translocation. Microduplication syndromes are often associated with a milder and more variable phenotype, leading to an ascertainment bias in grouping patient for molecular karyotyping (Table 5.2).

For a more detailed description of these novel syndromes, readers are referred to other publications (Cooper et al. 2011; Slavotinek 2008).

5.3.4 Detection of Disease-Related Genes by Identification of Index Cases

In some instances, array-CGH may be helpful in the discovery of monogenic diseases. In well-characterised patient groups with suspected "chromosomal phenotypes" array-CGH was useful in the identification of index cases to localise disease-causing genes. This could involve genes located at the translocation breakpoint or alterations of dosage-sensitive genes in combination with the detection of point mutations (Unger et al. 2008; Vissers et al. 2004). In even rarer instances, an inherited chromosomal deletion from an otherwise healthy parent

Table 5.1 List of microdeletion syndromes identified by molecular karyotyping approaches

Microdeletion syndrome (genomic position)	Size (Mb)	Phenotype	Reference
1q41–1q42	2.72–9.07	Variable, with DD, DM, MR, Seizures, midline defect, Fryns syndrome	Shaffer et al. (2007)
2p15	0.57–5.7	MR, autistic features, microcephaly, structural brain anomalies, kidney anomalies	Rajcan-Separovic et al. (2007)
4q21	0.37	DD, muscular hypotonia, severe MR, growth restriction, speech delay, distinctive face features, variable brain anomalies	Bonnet et al. (2010)
9q22.3	6.5	MR, psychomotor delay, hyperactivity, macrocephaly, overgrowth, trigonocephaly	Redon et al. (2006a)
15q13.3	1.5	MR, seizures, variable digital and facial dysmorphism	Sharp et al. (2008)
15q24	1.7–3.9	Growth retardation, microcephaly, digital abnormalities, loose connective tissue, facial dysmorphism	Sharp et al. (2007)
16p11.2–p12.12	7.1–8.7	DD, MR, hypotonia, hand and foot anomalies, short stature, facial dysmorphism	Ballif et al. (2007)
17q21.34	0.6	MR, hypotonia, characteristic facial dysmorphism	Koolen et al. (2006)
20q13.13–q13.2	1.3–2.8	Features of Okihiro syndrome with DD and overlapping features with CHARGE syndrome	Borozdin et al. (2007)
21q22.12	0.7	DD, MR, DM, thrombocytopenia	Shinawi et al. (2008)

Table 5.2 List of microduplication syndromes identified by molecular karyotyping approaches

Microduplication syndrome (genomic position)	Size (Mb)	Phenotype	Reference
3q29	1.9	Variable phenotype with DD, DM, ophthalmological findings	Ballif et al. (2008); Goobie et al. (2008)
Xq28	0.2–2.2	Severe MR, spasticity, asymptomatic females with skewed chromosome X inactivation	Van Esch et al. (2005)
7q11.23 (reciprocal of Williams–Beuren syndrome)	~1.5	MR, language delay	Somerville et al. (2005)
15q24	1.7	DD, facial dysmorphism, digital and genital abnormalities	Kiholm Lund et al. (2008)
17p11.2	1.3–15.2	MR, hypotonia, autistic features, sleep apnoea, cardiovascular anomalies	Potocki et al. (2007)
22q11.2	~3	Mild, variable phenotype with familiar transmission	Ou et al. (2008)

may point to the presence of an inherited mutation in the second allele, resulting in an autosomal recessive phenotype (Lesnik Oberstein et al. 2006). In typical clinical settings, these findings are rather serendipitous, and the search for these monogenic (Mendelian) disorders is likely to be replaced by next-generation sequencing approaches (Majewski et al. 2011). Obviously, primary testing of these suspected syndromes should be sequencing-based. However, for mutation-negative cases, the use of targeted high-resolution array-CGH or MLPA could be beneficial (Emanuel and Saitta 2007; Szuhai et al. 2011).

5.3.5 *Detection of Mosaicism*

The analysis of mixed cell populations without the possibility of further separation of the different fraction is a daunting task for DNA-based testing. The accuracy of the detection of chromosomal mosaicism is heavily dependent of the number of analysed metaphases. With the molecular karyotyping approach, the detection of somatic mosaic conditions in which the abundance of minority cells is as low as 8% has been achieved (Ballif et al. 2006; Cheung et al. 2007; Szuhai et al. 2011). For prenatal diagnosis, the presence of maternal cell contamination may further complicate the detection (Shaffer et al. 2011). For tumour genome profiling, the presence of various degrees of normal cell contamination (stromal and inflammatory cells) in combination with increased tumour ploidy decreases the detection sensitivity of molecular karyotyping approaches (Kallioniemi 2008).

5.3.6 *Prenatal Screening*

Next to the already described inherent limitation of conventional karyotyping for prenatal testing (resolution, failure to culture cells), one of the most important issue is the time required to obtain results. For targeted approaches, such as trisomy 21 screening based on advanced maternal age in combination with serum markers and ultrasound test, several molecular approaches have been successfully applied. FISH, qPCR or MLPA-based detections of aneuploidies for chromosomes 13, 18, 21, X and Y have been adapted in clinical settings (Faas et al. 2011; Van Opstal et al. 2009). The molecular karyotyping approach can obviously detect all abnormalities discovered by G-banding or any whole-chromosome aberration. In addition, molecular karyotyping using high-resolution interrogation of the genome for copy number alterations facilitates the detection of previously undetected submicroscopic rearrangements and mosaic alterations (Coppinger et al. 2009; Faas et al. 2010; Shaffer et al. 2008). Interpretation of the detected genomic imbalances associated with well-established DD/MR or MCR is easier compared to detected CNVs with unknown clinical significance. Arrays designed with targeted reporters may in part overcome these diagnostic dilemmas (Coppinger et al. 2009; Shaffer and Bejjani 2011).

On the other hand, with the targeted approach, genomic loci with clinically relevant imbalances may be overlooked due to the very low recurrence rate and the great degree of genomic heterogeneity (see Sect. 5.3.1) (Cooper et al. 2011; Edelmann and Hirschhorn 2009; Lu et al. 2007).

5.3.7 *Molecular Karyotyping in Neoplasia*

Various types of chromosomal aberrations involving chromosomal copy number alterations (CNA) and structural aberrations (translocations, deletions, amplifications) are commonly found in human neoplasia. Solid tumours are often characterised by the presence of complex genome alterations, while haematological malignancies and some of the mesenchymal tumours are characterised by less complex genomic changes at chromosomal level. While most of the solid tumour-associated changes merely represent secondary changes indicating the presence of genomic instability, certain changes may help in the identification prognostic groups. A typical example is the homozygous loss CDKN2A/B loci associated with poor outcome in several tumour entities. Genomic changes might be characteristic for certain tumour entities, such as the high level of amplification of the 12q13 region containing the MDM2 gene locus in most of the dedifferentiated liposarcoma cases and can be used as an adjunct tool in establishing the histopathological diagnosis. The investigation of the ERBB2 genomic locus in breast carcinoma is important for the treatment as presence of amplification of the ERBB2 locus identifies patients eligible for adjuvant treatment options (Kallioniemi 2008).

Due to the complex nature of the samples used for testing, tumour samples are often heterogeneous containing admixtures of different clones with various ploidy statuses and contaminating normal stromal elements. Therefore, the interpretation of array result is more difficult and may require detection and decision algorithms different as compared to the ones used for constitutional genetic testing. The overall variation of reporter signals may heavily depend on the quality of the input material. DNA samples might be degraded due to necrosis present before sampling in tumour sample, and samples retrieved from archives are often heavily degraded, especially when dealing with formalin-fixed paraffin-embedded (FFPE) tissue-derived material which is the most commonly used tissue processing in routine diagnostic pathology. However, both array-CGH and SNP-array testing have been successfully applied on various samples after different fixation procedures.

The use of high-resolution array-CGH testing instead of FISH-based testing for the detection of a homozygous deletion of theCDKN2A/B region has been shown to be superior, particularly in cases with deletion smaller than the size of the FISH probes. Array-based testing replacing locus-specific FISH probes for the detection of deletion or amplification could simultaneously provide information about the copy number status of several thousand additional loci.

For the discovery of genomic alterations that are clinically relevant, the use of array platforms with high resolution on very well-selected patient groups is inevitable.

For example, the application of an SNP-array platform with 1.8 million reporter elements on Philadelphia translocation-positive (Ph+) ALL and blastic phase transformed Ph + CML resulted in the identification of a hemizygous deletion on chromosome 7 with various sizes involving the IKZF1gene. The high probe density at the IKZF1 locus was necessary to detect this alteration as the critical minimal region of the deletion involved Δ3–6 containing 7 informative SNP mapped to the region (Mullighan et al. 2008).

The importance of platform selection and sample quality in array testing is essential for good results. For the identification of a known deletion or gain, array platforms with lower resolution can be used, while for discovery purposes, array platforms with the highest resolution should be selected (Albertson and Pinkel 2003; Kallioniemi 2008).

5.4 Discussion

Recent advances in molecular cytogenetic techniques have provided new tools to detect chromosomal rearrangements and have provided insight into the complexity of the human genome architecture. Molecular karyotyping using either array-CGH or SNP-array techniques are widely used for the detection of chromosomal copy number alterations. The introduction of array-based testing has led to a higher rate of detection of clinically relevant chromosomal abnormalities and resulted in the description of new chromosomal disorders (Cooper et al. 2011; Slavotinek 2008).

In addition, a large number of studies employing array-based technology in cancer have been published and have demonstrated the detection of copy number aberrations in various tumour types. These investigations lead to the identification of tumour-specific genetic alterations and genome changes which can be linked to disease progression, therapy response or prognosis (Kallioniemi 2008).

None of the molecular karyotyping approaches using genome-wide screening tool permits the detection of balanced chromosomal rearrangements such as translocation, inversion and insertion or the detection of polyploidy. The result obtained is a quantitative measurement of the abundance of genomic loci; however, it remains uninformative about the structural relation of the involved region. For example, for the full characterisation of a duplicated region, the FISH with probes covering the involved locus is necessary to discriminate a tandem duplication from an unbalanced insertion (Neill et al. 2011).

Genome-wide screening with microarrays with increasing resolution has great impact in the interpretation of results, especially when these tests are used in the context of prenatal testing but also for other applications. The development of assays with increased resolution at chromosomal loci with already known clinical relevance (interrogation of regions involved in microdeletion/microduplication syndromes and subtelomeric regions) in combination with a lower resolution overall chromosomal coverage avoiding known CNVs may be preferred in some instances . Understandably, any assay with higher resolution may uncover regions undetected by earlier formats with lower resolution. To study well-defined patient groups that have not yielded satisfactory results with other approaches such as MLPA or targeted gene sequencing, the use of dedicated arrays may improve the detection sensitivity for intragenic deletions down to a few kilobases.

The overall advantage of molecular karyotyping compared to classical karyotyping resulted in early recommendations to use it as an adjunct tool (Vermeesch et al. 2005) followed a few years later by recommendations to its implementation as a first-tier diagnostic tool (Miller et al. 2010). In several laboratories, molecular karyotyping has indeed been implemented as a standard test replacing karyotyping of G-banded chromosomes. Despite the increasing resolution, the detection rate of abnormalities has not increased linearly, pointing to mechanisms that (1) leave the overall genome balanced, (2) involve an epigenetic change or (3) are due to point mutations or small base substitutions. Recent advances related to sequencing technologies (next-generation sequencing: NGS) enable quick and affordable analysis of the whole genome. Dedicated sequencing approaches interrogating the coding region of the genome, also known as "exome sequencing" (Majewski et al. 2011), may present a reliable, combined mutation and copy number screening tool. Undoubtedly, the use of sequencing-based approaches will facilitate the identification of disease-causing genomic alteration associated with MR/DD and MCR in a significant proportion of the cases unresolved by molecular karyotyping approaches (Vissers et al. 2010).

References

Albertson DG, Pinkel D (2003) Genomic microarrays in human genetic disease and cancer. Hum Mol Genet 12(Spec No 2):R145–R152

Ballif BC, Rorem EA, Sundin K, Lincicum M, Gaskin S, Coppinger J, Kashork CD, Shaffer LG, Bejjani BA (2006) Detection of low-level mosaicism by array CGH in routine diagnostic specimens. Am J Med Genet A 140:2757–2767

Ballif BC, Hornor SA, Jenkins E, Madan-Khetarpal S, Surti U, Jackson KE, Asamoah A, Brock PL, Gowans GC et al (2007) Discovery of a previously unrecognized microdeletion syndrome of 16p11.2-p12.2. Nat Genet 39:1071–1073

Ballif BC, Theisen A, Coppinger J, Gowans GC, Hersh JH, Madan-Khetarpal S, Schmidt KR, Tervo R, Escobar LF et al (2008) Expanding the clinical phenotype of the 3q29 microdeletion syndrome and characterization of the reciprocal microduplication. Mol Cytogenet 1:8

Bonnet C, Andrieux J, Beri-Dexheimer M, Leheup B, Boute O, Manouvrier S, Delobel B, Copin H, Receveur A et al (2010) Microdeletion at chromosome 4q21 defines a new emerging syndrome with marked growth restriction, mental retardation and absent or severely delayed speech. J Med Genet 47:377–384

Borozdin W, Graham JM Jr, Bohm D, Bamshad MJ, Spranger S, Burke L, Leipoldt M, Kohlhase J (2007) Multigene deletions on chromosome 20q13.13-q13.2 including SALL4 result in an expanded phenotype of Okihiro syndrome plus developmental delay. Hum Mutat 28:830

Cheung SW, Shaw CA, Scott DA, Patel A, Sahoo T, Bacino CA, Pursley A, Li J, Erickson R et al (2007) Microarray-based CGH detects chromosomal mosaicism not revealed by conventional cytogenetics. Am J Med Genet A 143A:1679–1686

Cooper GM, Coe BP, Girirajan S, Rosenfeld JA, Vu TH, Baker C, Williams C, Stalker H, Hamid R et al (2011) A copy number variation morbidity map of developmental delay. Nat Genet 43 (9):838–846

Coppinger J, Alliman S, Lamb AN, Torchia BS, Bejjani BA, Shaffer LG (2009) Whole-genome microarray analysis in prenatal specimens identifies clinically significant chromosome alterations without increase in results of unclear significance compared to targeted microarray. Prenat Diagn 29:1156–1166

Edelmann L, Hirschhorn K (2009) Clinical utility of array CGH for the detection of chromosomal imbalances associated with mental retardation and multiple congenital anomalies. Ann NY Acad Sci 1151:157–166

Eichler EE, Nickerson DA, Altshuler D, Bowcock AM, Brooks LD, Carter NP, Church DM, Felsenfeld A, Guyer M et al (2007) Completing the map of human genetic variation. Nature 447:161–165

Emanuel BS, Saitta SC (2007) From microscopes to microarrays: dissecting recurrent chromosomal rearrangements. Nat Rev Genet 8:869–883

Faas BH, van der Burgt I, Kooper AJ, Pfundt R, Hehir-Kwa JY, Smits AP, de Leeuw N (2010) Identification of clinically significant, submicroscopic chromosome alterations and UPD in fetuses with ultrasound anomalies using genome-wide 250k SNP array analysis. J Med Genet 47:586–594

Faas BH, Cirigliano V, Bui TH (2011) Rapid methods for targeted prenatal diagnosis of common chromosome aneuploidies. Semin Fetal Neonatal Med 16:81–87

Flint J, Knight S (2003) The use of telomere probes to investigate submicroscopic rearrangements associated with mental retardation. Curr Opin Genet Dev 13:310–316

Flint J, Wilkie AO, Buckle VJ, Winter RM, Holland AJ, McDermid HE (1995) The detection of subtelomeric chromosomal rearrangements in idiopathic mental retardation. Nat Genet 9:132–140

Freeman JL, Perry GH, Feuk L, Redon R, McCarroll SA, Altshuler DM, Aburatani H, Jones KW, Tyler-Smith C et al (2006) Copy number variation: new insights in genome diversity. Genome Res 16:949–961

Goobie S, Knijnenburg J, Fitzpatrick D, Sharkey FH, Lionel AC, Marshall CR, Azam T, Shago M, Chong K et al (2008) Molecular and clinical characterization of de novo and familial cases with microduplication 3q29: guidelines for copy number variation case reporting. Cytogenet Genome Res 123:65–78

Kallioniemi A (2008) CGH microarrays and cancer. Curr Opin Biotechnol 19:36–40

Kallioniemi A, Kallioniemi OP, Sudar D, Rutovitz D, Gray JW, Waldman F, Pinkel D (1992) Comparative genomic hybridization for molecular cytogenetic analysis of solid tumors. Science 258:818–821

Kiholm Lund AB, Hove HD, Kirchhoff M (2008) A 15q24 microduplication, reciprocal to the recently described 15q24 microdeletion, in a boy sharing clinical features with 15q24 microdeletion syndrome patients. Eur J Med Genet 51:520–526

Kirchhoff M, Rose H, Lundsteen C (2001) High resolution comparative genomic hybridisation in clinical cytogenetics. J Med Genet 38:740–744

Knijnenburg J, Oberstein SA, Frei K, Lucas T, Gijsbers AC, Ruivenkamp CA, Tanke HJ, Szuhai K (2009) A homozygous deletion of a normal variation locus in a patient with hearing loss from non-consanguineous parents. J Med Genet 46:412–417

Koolen DA, Vissers LE, Pfundt R, Knight SJ, Regan R, Kooy RF, Reyneirs E, Romano C et al (2006) A new chromosome 17q21.31 microdeletion syndrome associated with a common inversion polymorphism. Nat Genet 38:999–1001

Lee C, Iafrate AJ, Brothman AR (2007) Copy number variations and clinical cytogenetic diagnosis of constitutional disorders. Nat Genet 39:S48–S54

Lesnik Oberstein SA, Kriek M, White SJ, Kalf ME, Szuhai K, den Dunnen JT, Breuning MH, Hennekam RC (2006) Peters plus syndrome is caused by mutations in B3GALTL, a putative glycosyltransferase. Am J Hum Genet 79:562–566

Lu X, Shaw CA, Patel A, Li J, Cooper ML, Wells WR, Sullivan CM, Sahoo T, Yatsenko SA et al (2007) Clinical implementation of chromosomal microarray analysis: summary of 2513 postnatal cases. PLoS One 2:e327

Majewski J, Schwartzentruber J, Lalonde E, Montpetit A, Jabado N (2011) What can exome sequencing do for you? J Med Genet 48:580–589

Miller DT, Adam MP, Aradhya S, Biesecker LG, Brothman AR, Carter NP, Church DM, Crolla JA, Eichler EE et al (2010) Consensus statement: chromosomal microarray is a first-tier clinical diagnostic test for individuals with developmental disabilities or congenital anomalies. Am J Hum Genet 86:749–764

Mullighan CG, Miller CB, Radtke I, Phillips LA, Dalton J, Ma J, White D, Hughes TP, Le Beau MM et al (2008) BCR-ABL1 lymphoblastic leukaemia is characterized by the deletion of Ikaros. Nature 453:110–114

Neill NJ, Ballif BC, Lamb AN, Parikh S, Ravnan JB, Schultz RA, Torchia BS, Rosenfeld JA, Shaffer LG (2011) Recurrence, submicroscopic complexity, and potential clinical relevance of copy gains detected by array CGH that are shown to be unbalanced insertions by FISH. Genome Res 21:535–544

Ou Z, Berg JS, Yonath H, Enciso VB, Miller DT, Picker J, Lenzi T, Keegan CE, Sutton VR et al (2008) Microduplications of 22q11.2 are frequently inherited and are associated with variable phenotypes. Genet Med 10:267–277

Pinkel D, Segraves R, Sudar D, Clark S, Poole I, Kowbel D, Collins C, Kuo WL, Chen C et al (1998) High resolution analysis of DNA copy number variation using comparative genomic hybridization to microarrays. Nat Genet 20:207–211

Potocki L, Bi W, Treadwell-Deering D, Carvalho CM, Eifert A, Friedman EM, Glaze D, Krull K, Lee JA et al (2007) Characterization of Potocki-Lupski syndrome (dup(17)(p11.2p11.2)) and delineation of a dosage-sensitive critical interval that can convey an autism phenotype. Am J Hum Genet 80:633–649

Raap AK, van der Burg MJ, Knijnenburg J, Meershoek E, Rosenberg C, Gray JW, Wiegant J, Hodgson JG, Tanke HJ (2004) Array comparative genomic hybridization with cyanin cis-platinum-labeled DNAs. Biotechniques 37:130–134

Rajcan-Separovic E, Harvard C, Liu X, McGillivray B, Hall JG, Qiao Y, Hurlburt J, Hildebrand J, Mickelson EC et al (2007) Clinical and molecular cytogenetic characterisation of a newly recognised microdeletion syndrome involving 2p15-16.1. J Med Genet 44:269–276

Redon R, Baujat G, Sanlaville D, Le Merrer M, Vekemans M, Munnich A, Carter NP, Cormier-Daire V, Colleaux L (2006a) Interstitial 9q22.3 microdeletion: clinical and molecular characterisation of a newly recognised overgrowth syndrome. Eur J Hum Genet 14:759–767

Redon R, Ishikawa S, Fitch KR, Feuk L, Perry GH, Andrews TD, Fiegler H, Shapero MH, Carson AR et al (2006b) Global variation in copy number in the human genome. Nature 444:444–454

Sebat J, Lakshmi B, Malhotra D, Troge J, Lese-Martin C, Walsh T, Yamrom B, Yoon S, Krasnitz A et al (2007) Strong association of de novo copy number mutations with autism. Science 316:445–449

Shaffer LG, Bejjani BA (2011) Development of new postnatal diagnostic methods for chromosome disorders. Semin Fetal Neonatal Med 16:114–118

Shaffer LG, Bejjani BA, Torchia B, Kirkpatrick S, Coppinger J, Ballif BC (2007) The identification of microdeletion syndromes and other chromosome abnormalities: cytogenetic methods of the past, new technologies for the future. Am J Med Genet C Semin Med Genet 145C:335–345

Shaffer LG, Coppinger J, Alliman S, Torchia BA, Theisen A, Ballif BC, Bejjani BA (2008) Comparison of microarray-based detection rates for cytogenetic abnormalities in prenatal and neonatal specimens. Prenat Diagn 28:789–795

Shaffer LG, Coppinger J, Morton SA, Alliman S, Burleson J, Traylor R, Walker C, Byerly S, Lamb AN et al (2011) The development of a rapid assay for prenatal testing of common aneuploidies and microdeletion syndromes. Prenat Diagn 31:778–787

Sharp AJ, Selzer RR, Veltman JA, Gimelli S, Gimelli G, Striano P, Coppola A, Regan R, Price SM et al (2007) Characterization of a recurrent 15q24 microdeletion syndrome. Hum Mol Genet 16:567–572

Sharp AJ, Mefford HC, Li K, Baker C, Skinner C, Stevenson RE, Schroer RJ, Novara F, De Gregori M et al (2008) A recurrent 15q13.3 microdeletion syndrome associated with mental retardation and seizures. Nat Genet 40:322–328

Shaw-Smith C, Redon R, Rickman L, Rio M, Willatt L, Fiegler H, Firth H, Sanlaville D, Winter R et al (2004) Microarray based comparative genomic hybridisation (array-CGH) detects submicroscopic chromosomal deletions and duplications in patients with learning disability/mental retardation and dysmorphic features. J Med Genet 41:241–248

Shinawi M, Cheung SW (2008) The array CGH and its clinical applications. Drug Discov Today 13:760–770

Shinawi M, Erez A, Shardy DL, Lee B, Naeem R, Weissenberger G, Chinault AC, Cheung SW, Plon SE (2008) Syndromic thrombocytopenia and predisposition to acute myelogenous leukemia caused by constitutional microdeletions on chromosome 21q. Blood 112:1042–1047

Slavotinek AM (2008) Novel microdeletion syndromes detected by chromosome microarrays. Hum Genet 124:1–17

Solinas-Toldo S, Lampel S, Stilgenbauer S, Nickolenko J, Benner A, Dohner H, Cremer T, Lichter P (1997) Matrix-based comparative genomic hybridization: biochips to screen for genomic imbalances. Genes Chromosomes Cancer 20:399–407

Somerville MJ, Mervis CB, Young EJ, Seo EJ, Bamforth S, Peregrine E, Loo W, Lilley M et al (2005) Severe expressive-language delay related to duplication of the Williams-Beuren locus. N Engl J Med 353:1694–1701

Speicher MR, Carter NP (2005) The new cytogenetics: blurring the boundaries with molecular biology. Nat Rev Genet 6:782–792

Stankiewicz P, Beaudet AL (2007) Use of array CGH in the evaluation of dysmorphology, malformations, developmental delay, and idiopathic mental retardation. Curr Opin Genet Dev 17:182–192

Szuhai K, Jennes I, Bovee JVMG, Wiwegar M, Wuyts W, Hogendoorn PCW (2011) Tiling resolution array-CGH shows that somatic mosaic deletion of the EXT gene is causative in EXT gene mutation negative multiple osteochondromas patients. Hum Mutat 32: E2036–E2049

Unger S, Bohm D, Kaiser FJ, Kaulfuss S, Borozdin W, Buiting K, Burfeind P, Bohm J, Barrionuevo F et al (2008) Mutations in the cyclin family member FAM58A cause an X-linked dominant disorder characterized by syndactyly, telecanthus and anogenital and renal malformations. Nat Genet 40:287–289

Van Esch H, Bauters M, Ignatius J, Jansen M, Raynaud M, Hollanders K, Lugtenberg D, Bienvenu T, Jensen LR et al (2005) Duplication of the MECP2 region is a frequent cause of severe mental retardation and progressive neurological symptoms in males. Am J Hum Genet 77:442–453

Van Opstal D, Boter M, de Jong D, van den Berg C, Bruggenwirth HT, Wildschut HI, de Klein A, Galjaard RJ (2009) Rapid aneuploidy detection with multiplex ligation-dependent probe amplification: a prospective study of 4000 amniotic fluid samples. Eur J Hum Genet 17:112–121

Vermeesch JR, Melotte C, Froyen G, Van Vooren S, Dutta B, Maas N, Vermeulen S, Menten B, Speleman F et al (2005) Molecular karyotyping: array CGH quality criteria for constitutional genetic diagnosis. J Histochem Cytochem 53:413–422

Vissers LE, De Vries BB, Osoegawa K, Janssen IM, Feuth T, Choy CO, Straatman H, Van d V, Huys EH et al (2003) Array-based comparative genomic hybridization for the genomewide detection of submicroscopic chromosomal abnormalities. Am J Hum Genet 73:1261–1270

Vissers LE, van Ravenswaaij CM, Admiraal R, Hurst JA, De Vries BB, Janssen IM, van der Vliet WA, Huys EH, de Jong PJ et al (2004) Mutations in a new member of the chromodomain gene family cause CHARGE syndrome. Nat Genet 36:955–957

Vissers LE, de Ligt J, Gilissen C, Janssen I, Steehouwer M, de Vries P, van Lier B, Arts P, Wieskamp N et al (2010) A de novo paradigm for mental retardation. Nat Genet 42:1109–1112

Part II
Protein (Antigen and Antibody) and Peptide Arrays

Chapter 6
Protein Microarrays and Their Potential Clinical Applications in the Era of Personalized Medicine

Hong Zhang and Steven Pelech

6.1 Introduction

Despite tremendous advances in our understanding of the molecular basis of human diseases and enormous improvements in clinical technologies, substantial progress in diagnosis and treatment has yet to materialize. The disparity can be at least in part attributed to the complexity and heterogeneity of human diseases with respect to the underlying molecular mechanisms, which have become increasingly recognized as a result of decades of research. Cancers, for example, arise from the accumulations of alterations of signaling pathways involved in various physiological processes such as cell growth, proliferation, apoptosis, and angiogenesis (Hanahan and Weinberg 2011). The existing clinical parameters employed for diagnosis and prognosis have proven to be inadequate especially in dealing with complex diseases like cancer. Indeed, many of the most common ailments, including diabetes and neurological and immunological disorders, appear to arise from a variety of multigenetic causes.

Delineation of disease-linked alterations at the molecular level through biomolecular profiling of biological specimens will provide a more comprehensive picture of deregulated cell signaling cascades. This will facilitate the identification of panels of biomarkers in a disease- and patient-specific manner for better diagnosis, prognosis, patient stratifications, as well as patient-specific treatment. In the foreseeable future, biomolecular profiling is expected to become an integrated part of the routine clinical practice, supplementary to traditional pathological examinations for disease diagnosis, prognosis, and prediction of drug response in the era of

H. Zhang • S. Pelech (✉)
Kinexus Bioinformatics Corporation, Vancouver, BC, Canada
e-mail: hzhang@kinexus.ca; spelech@kinexus.ca

B. Jordan (ed.), *Microarrays in Diagnostics and Biomarker Development*,
DOI 10.1007/978-3-642-28203-4_6,

personalized medicine. Recent advances of high-throughput and high-content systems biology technologies have enabled large-scale analysis of hundreds and even thousands of cell signaling molecules in a single assay. Among them, microarrays have arisen to be an effective and versatile platform that would allow the analyses of complex clinical samples in a miniaturized and parallel fashion.

The completion of the human genome sequencing project early in last decade drove the development of DNA microarrays, which allows gene expression profiling on a genome-wide scale. They have since become powerful tools for identification of gene expression patterns that are potentially associated with the state and progression of a disease (Baehner et al. 2011). While the implementation of DNA microarrays in the analysis of clinical samples has led to discoveries of gene regulation linked to specific diseases, genome profiling is limited in providing insight into the dynamics of cellular signaling networks with respect to the levels, subcellular locations, and functional activity states of proteins or their specific interactions. Such posttranslational information is essential for the understanding of molecular events leading to malignant transformation. Nevertheless, many years of technological innovations in the area of DNA microarrays have paved the way for the emergence and development of protein microarrays.

In a broad sense, protein microarrays are made up of immobilized "bait" molecular probes on a planar surface, which can be either homogeneous or heterogeneous in nature, including antibodies, recombinant proteins, peptides, or complex protein samples. While several variants on protein microarrays have been reported, they are typically comprised of hundreds or even thousands of protein or peptide molecules immobilized on modified array substrates in a spatial addressable format, analogous to DNA microarrays. They can be broadly classified into two main categories based on the types of molecules arrayed. The forward-phase protein microarrays consist of an array of capture ligands such as antibodies or antibody-like molecules that allow parallel detection of a large number of analytes present in a complex biological sample such as body fluids and crude protein lysates. The most common types of forward-phase protein microarrays include antibody microarrays, designed for identifying and quantifying proteins of interest, and recombinant protein microarrays, which are intended for function studies of interactions between immobilized proteins and other proteins, peptides, or compounds. In contrast, reverse-phase protein microarrays (RPPMs), as an orthogonal approach to antibody microarrays, feature a multitude of protein lysates as individual spots on the array surface, each of which might represent as much as the entire proteome of a biological sample. The RPPM assays allow the detection and quantification of select target proteins recognized by the probing antibodies across a large number of samples.

In this chapter, we will describe recent technological developments pertaining to the field of protein microarrays, antibody microarrays, and RPPMs in particular. A representative review of some of the latest work in this field in relation to clinical implementations in the context of personalized medicine is provided. Potential and challenges for clinical applications of the technologies are also discussed.

6.2 Recent Development of Protein Microarray Technology

The higher sensitivity of the microarray-based analyses afforded by assay miniaturization stems from the improvement of signal-to-noise ratios and decreased incubation time due to shorter diffusion distances between probes and their targets, based on the principle described by Ekins and Chu (1991). Although the concept of miniaturized immunoassays was introduced more than two decades ago, it was not until 10 years after the advent of DNA microarrays that the first antibody-based arrays became described (Knezevic et al. 2001). The lag in the development of protein microarrays was largely attributed to the complex nature of proteomes as a result of alternative splicing, posttranslational modifications, and protein–protein interactions. The human proteome arises from approximately 23,000 genes, which are predicted to encode over a million protein species when various spliced and covalently modified forms are taken into account. Thus, producing protein microarrays turned out to be much more challenging than simply adopting basic concepts and methodologies from DNA microarrays. This may account for the slow development and moderate success in the area of protein microarrays that have been witnessed so far, in spite of initial high expectations.

Nevertheless, the technologies pertaining to protein microarrays have been evolving in a steady pace in the past decade from the proof-of-concept to off-the-shelf products or services that are available commercially (see Sect. 6.2.1). This is a consequence of cross-disciplinary efforts in tackling a wide range of technical issues, including production of affinity probes, modifications of array surfaces for protein immobilization, optimization of sample preparations, and signal detection. While disease biology is still the main focus of the vast majority of protein microarray applications, there has been somewhat limited success of protein microarrays in clinical settings even in recent years as evidenced by relatively few publications in the field.

6.2.1 Antibody Microarrays

Antibody microarrays are miniaturized analytical platforms fabricated by arraying minute amounts of individual antibodies or antibody-like molecules onto a solid support, which can be either a planar surface, like glass slides, or microspheres made, for example, from polystyrene. Multiplexing capability of microsphere-based antibody microarrays or "liquid" arrays is achieved by immobilizing different capture molecules on microspheres coded with distinct flourophores emitting at different wavelengths. Identification of target binding is achieved by sorting microspheres based on the flow cytometric principle according to the combination of internal fluorescent signals from the beads and target-specific signals. The limitation of multiplexing capability, the requirement of larger sample amounts, as well as the need for specialized analytical equipment such as those developed by

Luminex (Austin TX, http://www.luminexcorp.com) have limited their broad applications. In comparison, planar protein microarrays only need very small amounts of biological samples. They appear in one of two basic formats. Some are based on the ELISA sandwich setup that relies on two separate antibodies to capture and to detect, respectively, a target protein. Despite higher sensitivity and specificity, the cross-reactivity inherent to antibodies significantly limits the multiplexing capability of this type of antibody microarrays. Other limiting factors include the availability of suitable pairs of capture and detection antibodies and the number of different internally tagged microspheres that can be deployed for analysis of a single sample. More and more antibody microarrays are designed in such a way that only a single antibody in each spot is utilized to capture its target protein that is pre-labeled with either fluorescent dyes or haptens. The multiplexing capability in this case is only limited by the availability of antibodies. Thus, single-antibody planar protein microarrays are the most commonly used platform today and likely in the foreseeable future.

The interest in the antibody microarray technology surged around the middle of the last decade, as evidenced by the emergence of a number of companies carrying a variety of antibody microarray-related products. The low data reliability resulting from poor product quality control with the early generations of antibody microarrays manufactured by simply adopting DNA microarray concepts, methodologies, and materials has exerted a negative impact on the confidence of investigators, casting doubt on their value in research. In the past couple of years, the landscape of the antibody microarray field has undergone some dramatic changes, including technological improvement on the production of antibodies and array substrates, as well as the optimization of printing, probing, and signal detection procedures that are specific for protein microarrays. On the one hand, the content of antibody microarrays has expanded from low (several dozens), in the early days, to median density (around 1,000 probes) as more and more antibodies become available. On the other hand, emphasis has been placed on the data validations using independent approaches such as immunoblotting and ELISA. A recent survey of the antibody microarray market revealed a list of less than a dozen of companies still offering antibody microarrays as either products or services. Information related to these products or services can be found in Table 6.1.

A key limiting factor for the development of antibody microarrays is the availability of high-quality antibodies that are suitable for microarray applications. Unlike DNA or oligonucleotide probes employed in DNA microarrays, antibodies and other affinity probes required for antibody microarrays cannot be produced with predicted affinity and specificity; this can be at least in part attributed to the complexity of protein–antibody interactions. Therefore, to generate a comprehensive collection of affinity probes that are well characterized and that can target as many of individual target proteins and their specific modified forms as possible with minimal cross-reactivities for off-target proteins is the most important task in the development of antibody microarrays.

Early generations of antibody microarrays consisted mainly of antibodies targeting cell surface antigens and cytokines, due to their widespread availability

Table 6.1 Commercial antibody microarrays and their specifications

Vendor	Item	Product/service	Capture antibodies
Arrayit	1	PlasmaScan Antibody Microarrays	80 Monoclonal antibodies
Clontech	2	Antibody Microarray 500	500 Monoclonal antibodies
Full Moon BioSystems/ Eurogentec	3	Antibody Microarrays	1,318 Polyclonal antibodies
Hypromatrix	4	Signal Transduction Antibody Array	400 Polyclonal antibodies
Kinexus Bioinformatics	5	Kinex™ Antibody Microarray Services	810 Polyclonal and monoclonal antibodies
Panomics/Affymetrix	6	Angiogenesis Antibody Arrays	36 Monoclonal antibodies
RayBiotech	7	RayBio® Human Cytokine Array	Over 200 Polyclonal and monoclonal antibodies
	8	RayBio® Human RTK Phosphorylation Antibody Array	71 Polyclonal and monoclonal antibodies
R&D Systems	9	Proteome Profiler Antibody Arrays	Up to 119 polyclonal antibodies
	10	Proteome Profiler 96 Antibody Arrays	Up to 16 antibodies per well
Sigma-Aldrich	11	Panorama Antibody Microarray XP725	725 Polyclonal antibodies
Spring Bioscience	12	AMS700 Master Antibody Microarrays	656 Polyclonal and monoclonal antibodies

Vendor	Item	Array substrate	Target proteins	P-site profiling
Arrayit	1	Glass slide	Human serum proteins	No
Clontech	2	Glass slide	Signaling proteins	No
Full Moon BioSystems/ Eurogentec	3	Glass slide	Signaling proteins	Yes
Hypromatrix	4	Nitrocellulose membrane	Signaling proteins	Few
Kinexus Bioinformatics	5	Glass slide	Signaling proteins	Yes
Panomics/Affymetrix	6	Nitrocellulose membrane	Cytokines	No
RayBiotech	7	Nitrocellulose membrane/ glass slide	Cytokines	No
	8	Nitrocellulose membrane/ glass slide	Receptor-tyrosine kinases	Yes
R&D Systems	9	Nitrocellulose membrane	Signaling proteins	Yes
	10	96-Well microtiter plate	Signaling proteins	Yes
Sigma-Aldrich	11	Glass slide	Signaling proteins	Few
Spring Bioscience	12	Glass slide	Signaling proteins	Few

Vendor	Item	Assay setup	Detection
Arrayit	1	Single-antibody assay	Direct labeling-2 color
Clontech	2	Single-antibody assay	Direct labeling-2 color
Full Moon BioSystems/Eurogentec	3	Single-antibody assay	Direct labeling-1 color
Hypromatrix	4	Sandwich assay	Label-free, Chemiluminescence

(continued)

Table 6.1 (continued)

Vendor	Item	Assay setup	Detection
Kinexus Bioinformatics	5	Single-antibody assay	Direct labeling-1 color
Panomics/Affymetrix	6	Sandwich assay	Label-free, Chemiluminescence
RayBiotech	7	Sandwich assay	Chemiluminescence/ fluorescence
	8	Sandwich assay	Chemiluminescence/ fluorescence
R&D Systems	9	Sandwich assay	Chemiluminescence
	10	Sandwich assay	Chemiluminescence
Sigma-Aldrich	11	Single-antibody assay	Direct labeling-2 color
Spring Bioscience	12	Single-antibody assay	Direct labeling-1 color

and better characterization in ELISA-based immunoassays. In contrast, development of antibody microarrays that track intracellular signaling proteins has been lagging, as a result of the shortage of high-quality antibodies and difficulty in extracting low abundance target proteins from complex biological samples. Despite increasing number of antibodies available in recent years, antibodies for significant portion of the human proteome are still lacking, not to mention those targeting only a tiny fraction of the posttranslational modification forms of the proteins. It has been a common practice to spot almost any commercial antibodies, both monoclonal and polyclonal, that are available on antibody microarrays within the scientific community and the industry. The lack of detailed characterizations of antibodies specific for the antibody microarray application has raised more questions than answers in the studies where such antibody microarrays were utilized. The importance of application-specific characterization of antibodies was highlighted by the work carried out by Patrick Brown's group, which revealed that less than 20% of monoclonal antibodies arrayed on the array surfaces displayed the expected reactivity (Haab et al. 2001). This observation was also made later by Gavin MacBeath's group, when they showed that only about 5% of over 100 commercial monoclonal antibodies that they tested were suitable for microarray application (MacBeath 2002).

To further complicate the matter, increasing numbers of commercial antibodies supplied by an ever-growing number of antibody vendors focusing on a small number of popular protein targets, in combination with poor quality controls, have made identifying a quality antibody like finding a needle in a haystack. According to the listings of two popular antibody search engines, Antibody Resource Page (http://www.antibodyresource.com) and Antibody and Beyond (http://www.antibodybeyond.com), there are at least 350 antibody suppliers that are currently active around the world carrying more than half a million antibodies.

Thanks to the efforts of some major collaborative initiatives from the scientific community worldwide, the situation has been improved dramatically in the last few years. Projects aiming at systematic generation and characterization of antibodies with the ultimate goal of obtaining antibodies for each of the proteins in the proteome were launched, including the Antibody Resource database (http://www.antibodypedia.org), part of the HUPO's Human Antibody Initiative (HAI). Through its Human Protein Atlas Program, more than 12,000 monospecific polyclonal antibodies have been produced so far against close to 9,000 human proteins, which correspond to about −39% of all human protein-encoding genes. These antibodies have been extensively characterized by Western blotting as well as immunohistochemistry (IHC) on tissue microarrays. Other notables include the ProteomeBinders (http://www.proteomebinders.org), the Clinical Proteomic Reagents Resources within Clinical Proteomic Technologies Initiative for Cancer (CPTI) (http://www.proteomics.cancer.gov), and the Antibody Factory (http://www.antibody-factory.de). Few of these antibodies, however, target covalent modifications of proteins such as protein phosphorylation. At Kinexus Bioinformatics Corporation, we have recently launched a large-scale antibody development project with the assistance of the National Research Council of Canada Industrial Research Assistance Program, aimed at producing monospecific antibodies against hundreds of human phosphorylation sites identified through our bioinformatics programs, with a strong emphasis on functionally uncharacterized phosphosites using a novel multiplex immunization scheme coupled with epitope testing of the resulting sera with peptide arrays.

Given the astronomical number of proteins and their variants in the human proteome and the stringent requirement for antibody specificity, mass production of antibodies by animal immunization against every individual protein and its modified forms in the proteome presents a huge challenge both technically and financially. In recent years, large recombinant antibody phage display libraries have been explored as an alternative source of antibodies, as they tend to display a wide range of specificity, high on-chip stability, renewal ability, and small molecular sizes for allowing high-density spotting on the arrays. Recombinant antibody libraries have been constructed based on both single-chain Fv (scFv) and Fab fragments in which all fragments share the same stable molecular scaffold, resulting in very similar on-the-chip behavior while displaying unique specificities. However, progress in this front has been limited by the shortage of purified target proteins, which are essential for library screenings. Interesting, few if any phage display antibodies against specific target antibodies can be commercially sourced despite the availability of this technology for over a decade.

In addition, aptamers have been sought as affinity molecules in place of antibodies on antibody microarrays. Aptamers are short single-stranded oligonucleotides that have been selected for their specificity for proteins (Baldrich 2011). Thus, aptamer-based protein microarrays can be constructed on the exact same surfaces used by DNA microarrays or through in situ synthesis. However, despite their stability, the lower chemophysical diversity inherent to oligonucleotides compared to proteins may limit the chances for identifying aptamers specific for each protein target and favor strong interactions of low specificity between aptamers and target proteins. A review on the technical aspects and progresses on aptamer selections can be found in Chap. 8 of this book.

Early observations of reduced reactivity and/or specificity of antibodies after being spotted onto the array surfaces can be in part attributed to the use of array surface chemistries optimized for DNA microarrays. Immobilization of antibodies on such array surfaces may result in their partial denaturation, leading to the loss of activity and subsequent increase of nonspecific binding. Thus, it is very important to choose appropriate array surfaces for the success of antibody microarrays. A number of immobilization strategies have been evaluated extensively in several studies for preserving the on-chip activity, specificity, and stability of antibodies (Angenendt et al. 2003). They can be broadly classified into three major categories, including absorption, covalent conjugation, and affinity binding. Each has its own advantages and shortcomings, and the decision on which immobilization approach to use should be made on the basis of the assay conditions. For example, the Maxisorp™ black plastic microarray substrate from Nunc (http://www.nuncbrand.com) was a very popular surface for antibody microarray fabrications due to its success in ELISA application and its lower nonspecific binding. However, it suffers from unevenness of the background fluorescence, probably due to the heterogeneity in the flatness inherent to plastic-based surfaces (Zhang et al. unpublished). Its low binding capability for the antibodies spotted in very small areas on the arrays is evidenced by the common appearance of so-called comet tails as a result of spillover of spotted antibodies during array processing (Zhang et al. unpublished).

One of the key advantages of antibody microarrays is the ability to analyze thousands of analytes in parallel in crude protein lysates. Unlike most other high-throughput proteomics approaches such as those based on mass spectrometry, in which the complexity of samples needs to be reduced by fractionation and/or enrichment prior to analysis, sample pre-fractionation may not be required for antibody microarray analyses, due to the high specificity of many antibodies. The employment of antibodies with high sensitivity and specificity, as well as the use of fluorescent dyes to label protein molecules either directly or indirectly, permits the analysis of minute amounts of protein samples, i.e., 50 μg on the Kinex™ antibody microarrays from Kinexus, which is advantageous especially when working with precious biopsy samples that usually contain heterogeneous cell populations (Fig. 6.1). However, the use of non-denaturing conditions in antibody microarray assays retains the tertiary and quaternary structures of proteins. On the one hand, the retention of protein–protein complexes can mask epitopes on target proteins relevant for the recognition by antibodies on the arrays, leading to false-negative results. On the other hand, changes in the level of an interacting protein may also be misinterpreted to reflect alterations in the amount of the target protein, which would also result in false-positive results. Therefore, our group undertook a detailed investigation of the ability of various lysis buffers with different protein solubilizing strengths to unmask antigenic epitopes without compromising the interactions between target proteins and antibodies, but only came up with limited success. One of the most promising approaches we have been experimenting with for the sample preparation has been the generation of large peptide fragments following chemical digestions of intact protein species in lysates prior to protein

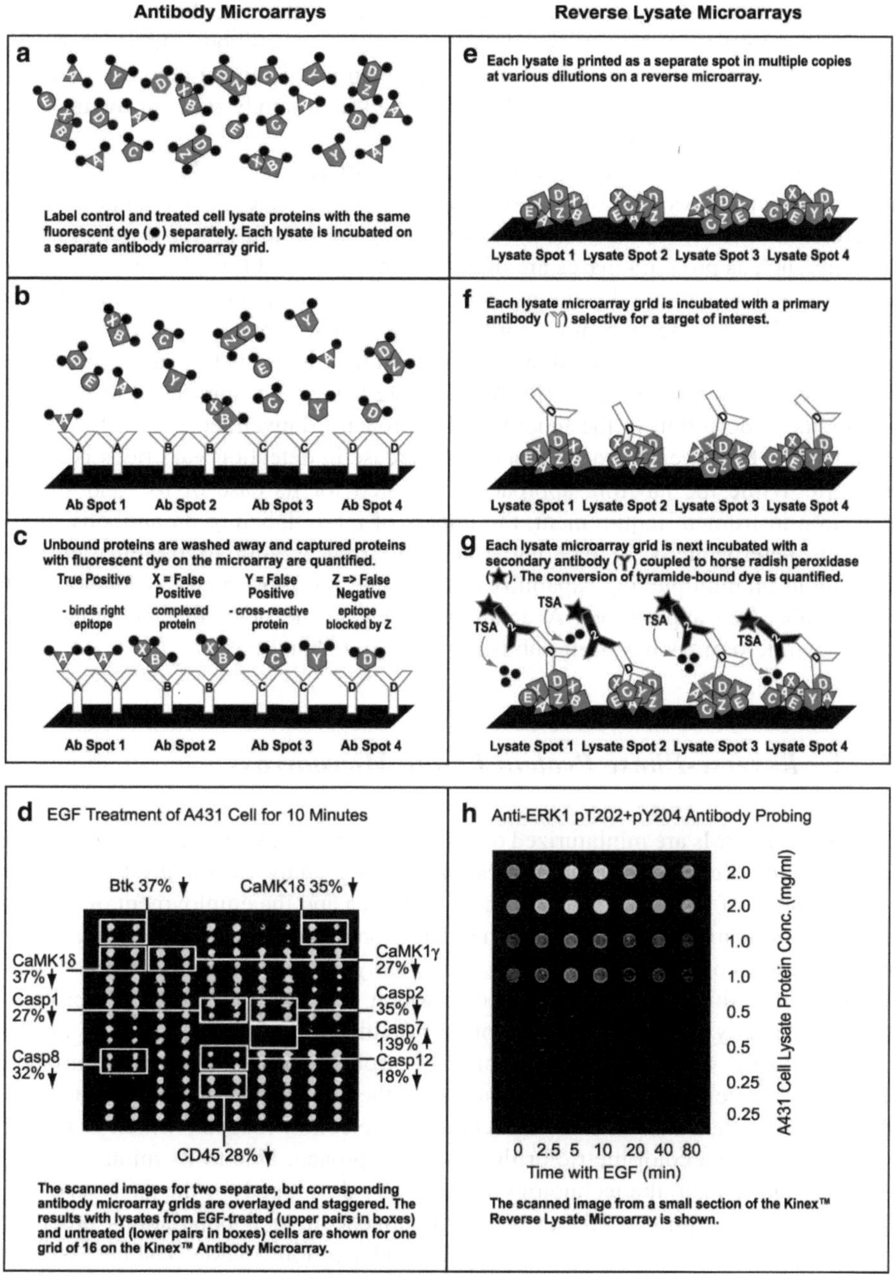

Fig. 6.1 Schematic diagrams of the Kinex™ antibody microarray (a–d) and protein lysate microarray (e–h) and their sample images

labeling for Kinex™ antibody microarray analyses. With prior chemical cleavage, we have observed a higher correlation between our antibody microarray and Western blot analysis results, indicating the potential of this strategy in reducing both false positives and false negatives observed in antibody microarray analyses (Zhang et al. unpublished).

Despite the high sensitivity and specificity offered by antibodies employed, antibody microarrays have often had limited success in detection and quantification of proteins of low abundance in complex biological samples. Signal transduction proteins such as protein kinases are often present at 100–1,000-fold lower levels in cell lysates than structural proteins and metabolic pathway enzymes. Label-based and label-free are the two most commonly used detection systems for antibody microarray analysis. Additional signal amplification steps such as those based on the rolling-circle amplification (RCA) principle have been introduced to improve the limits of detection in the label-based approach (Schweitzer et al. 2002; Shao et al. 2003). However, the accompanying increase in detection variations is also a concern. While the labeling approach is favored for its ease of use and simple detection instrument requirement, the impact of target labeling on antibody binding, the heterogeneity in labeling efficiency from protein to protein, and the limitations on multiplexing capability of the assays has motivated the development of the label-free approaches. A detailed review on the detection techniques for antibody microarrays has been published recently (Chandra et al. 2011).

6.2.2 Reverse-Phase Protein Lysate Microarrays

In essence, RPPMs are miniaturized dot Western blots, but with a much higher level of sensitivity than conventional Western blotting analyses, thanks to the high concentration of analytes located within a tiny area and the employment of various signal amplification strategies for signal detection. Minute amounts of hundreds or even thousands of protein samples are individually printed onto array surfaces often coated with a thin film of nitrocellulose, which displays extremely high protein-binding capacity (Fig. 6.1). Expression or modification of a protein across all the samples on the array can be monitored by probing the array with an antibody against the protein of interest. The binding of the antibody to its target proteins in the sample is then detected with a secondary detection reagent, which is further visualized using a colorimetric or fluorescent approach. Due to its minute sample amount requirement, the technology was initially developed for the analysis of the samples derived from small but highly homogeneous cell populations procured from biopsy samples using laser capture microdissection (LCM). The necessity of tumor cell enrichment using approaches like LCM was further demonstrated recently by the distinct differences observed for the expression and phosphorylation of 75 cell signaling proteins involved in cancer progression between the undissected whole tissue specimens and the matched microdissected samples on RPPM; this highlighted the utility of RPPMs for molecular profiling in the clinical setting (Silvestri et al. 2010). Protocols have also been developed and optimized to enable the extraction of proteins from formalin-fixation and paraffin-embedded (FFPE) tissue samples that may be suitable for the use on RPPMs, which may further pave the way for clinical application of this technology (Berg et al. 2010, 2011).

Similar to the situation with antibody microarrays, one of the limiting factors for developing RPPMs is the availability of highly specific antibodies. Despite the use of denatured protein samples similar to those used on Western blots, data derived from RPPMs and from Western blotting with the same set of samples and the same probing antibodies are not always in agreement with each other, despite that fact that Western blot analysis is routinely employed to screen and select antibodies for RPPM applications. Recently, Sevecka et al. (2011) screened 129 pan-specific and 254 posttranslational modification form-specific antibodies from commercial sources for their uses in the RPPM application, using a two-step system with the initial screening on a RPPM featuring 20 biological model systems followed by Western blots on select lysate samples. Less than 20% of the antibodies tested were validated to generate data consistent between RPPM and Western blots. There are multiple possible explanations for this discrepancy. In Western blotting, antibody cross-reactivity can be ignored as most of cross-reactive proteins can be separated from target proteins by gel electrophoresis on the basis of differences in size. In contrast, even minor cross-reactivity of the antibody may skew the quantification of spots on RPPMs. Based on the limited effect of protein phosphatases on cross-reactive signals from phospho-specific antibodies, Sevecka and coworkers concluded that the binding of off-target non-phosphoproteins is the main cause of signal compression on RPPMs. Therefore, it is imperative to characterize each antibody to be used in RPPMs in the same samples as those to be spotted onto the arrays to ensure its specificity for its cognate targets. While Western blotting remains as the main platform for testing antibody specificity prior to the use in RPPMs, an RNAi-based antibody validation approach involving RPPMs was recently proposed for selection of RPPM-compatible antibodies (Mannsperger et al. 2010).

In addition, it is also possible that much smaller amounts of proteins spotted on RPPMs place more stringent requirements on the assay's sensitivity when target proteins are present in low abundance in a mixture of tens of thousands of other protein molecules without any pre-fractionation or enrichment. To overcome this technical hurdle, a number of signal amplification strategies have been applied to improve signal detection on RPPMs. One of the common practices for signal detection was adopted from that routinely used in immunohistochemical analyses based on biotin–streptavidin–tyramide signal amplification reactions (DakoCytomation). Due to the narrow detection dynamic ranges inherent to the colorimetric detection, it is often necessary to spot protein samples in a serial dilution fashion in order to cover the linear detection range. In contrast, fluorescence-based signal amplification exhibits a much broader linear detection range, which permits better signal quantifications.

Furthermore, it was also speculated that the close proximity of protein molecules within the spot on RPPM might lead to the appearance of chimeric epitopes consisting of multiple distinct but spatially adjacent epitopes from different protein molecules that antibodies display cross-reactivity against (Sevecka et al. 2011).

However, whether or not this might be responsible for any inconsistencies observed between RPPM and Western blots needs further investigation.

6.3 Application of Antibody Microarrays for Disease Molecular Profiling and Biomarker Discovery

The popularity of antibody microarray technology for a wide range of clinical studies is unsurprising. Its capability for high-throughput profiling of a myriad of cell signaling proteins in a parallel fashion offers the potential to yield novel insights into the molecular mechanisms of pathological conditions, which in turn may lead to the identification of biomarkers for disease diagnosis and prognosis, patient stratifications, as well as prediction of therapeutic responses.

Despite technical challenges and concerns over its data accuracy and reproducibility, there has been no lack of attempts to use this technology to uncover molecular signatures of altered signaling pathways associated with diverse human pathological conditions using clinical samples such as serum, cerebrospinal fluid (CSF), and tissue biopsy specimens that are routinely accessible in the clinic. Despite the proof-of-concept nature, increasing numbers of antibody microarray-based clinical applications have been reported in recent years. While the technology is applicable to almost all pathological conditions, much of the emphasis has thus far been placed on biomarker identification for cancer diagnosis, prognosis, and therapeutic response prediction, due to the complex nature of the disease. Cancer has been described as a disease that arises from the multistep accumulation of genetic and epigenetic disorders, which lead to aberrant cellular functions including uncontrolled cellular proliferation, cell cycle deregulation, decreased cell death or apoptosis, differentiation inhibition, invasion, angiogenesis, and metastasis.

In one of the pioneering studies to explore the clinical potential of the antibody microarray, Knezevic et al. (2001) examined 368 cancer-related cellular signaling proteins in oral cavity tumor cell population procured by LCM on nitrocellulose-based antibody microarrays. Alterations of expression patterns of ribosomal S6 kinase (RSK) and progesterone receptor (PgR) within epithelial and stromal cells of oral cavity tumor were found to be correlated with tumor progression. Since then, the antibody microarray technology has been expanded to the investigation of a wide range of human cancers.

6.3.1 Prostate Cancer

Miller et al. (2003) used an antibody microarray comprising 184 monoclonal and polyclonal antibodies prepared on a hydrogel slide surface to profile protein expression in the serum samples obtained from prostate cancer patients and healthy controls. Five serum proteins were identified that were significantly different between the two groups, which were in agreement with ELISA measurements. By contrast, Fujita et al. (2008) investigated the cytokine profile in prostatic fluids obtained from cancerous prostate glands on the Cytokine Antibody Array from RayBio (Raybiotech, Norcross, GA). The overall profile of cytokine secretion uncovered correlated well with both cancer status and inflammation grade.

6.3.2 Breast Cancer

Using the BD Antibody Microarray 380 containing 378 monoclonal antibodies (Palo Alto, CA), Hudelist et al. (2004) compared the protein-expression pattern of malignant and adjacent normal breast tissue from a single breast cancer patient. Increased expression of proteins with broad physiological functions, such as casein kinase 1ε (CK1ε), MAP kinase kinase 7 (MKK7), CDC25C, annexin XI, p53, and eIF-4E, and a reduction in the level of 14-3-3ε in malignant breast tissue were observed when compared to the corresponding nonmalignant samples. Carlsson et al. (2008) constructed the first large-scale recombinant scFv antibody microarrays with 129 human recombinant scFv antibody fragments against 60 different proteins mainly involved in immunoregulation for profiling serum proteome in metastatic breast cancer patients. A panel of 11 altered proteins was uncovered, composed of nine nonredundant serum proteins that displayed a strong discriminating power for cancer versus healthy subjects with a 95% sensitivity and specificity, respectively, despite the fact that no correlation was found with the known clinical parameters including ER and PgR status, clinical stage, and histological features. A recent study utilizing our Kinex™ antibody microarray, which is capable of tracking the expression and phosphorylation levels of over 600 intracellular signaling proteins, indicated that Y-box binding protein-1 (YB-1) appeared to facilitate breast cancer carcinogenesis through the induction of several cell cycle-related proteins including LIM kinase 1/2 (LIMK1/2), which resulted in the blockage of cytokinesis and led to the appearance of the cells harboring with numerical and structural chromosomal abnormalities as well as human EGF receptor 2 (HER2/ErbB2/Neu) amplification (Davies et al. 2011).

6.3.3 Pancreatic Cancer

Orchekowski et al. (2005) applied a 92-antibody microarray to profile molecular alterations in blood in relation to pancreatic cancer using serum samples from pancreatic cancer patients, patients with benign pancreatic disease, and healthy control subjects. The proteins that displayed reproducibly different binding levels between the patient classes included those involved in inflammation, immune response, leakage of cell breakdown products, and possibly altered vitamin K usage or glucose regulation. The authors showed that the patient classification accuracies were significantly improved over those achieved using individual antibodies; with 90% sensitivity and 93% specificity in discriminating the cancer samples from the healthy control samples and with 92% sensitivity and 81% specificity in separating the cancer samples from the benign disease samples. Ingvarsson et al. (2008) employed a recombinant scFv antibody microarray to identify protein signatures associated with pancreatic adenocarcinoma. A panel of 19 serum proteins was identified that discriminated between cancer patients and healthy subjects and was correlated with a subset of cancer patients having a life expectancy of greater than 12 months.

6.3.4 Lung Cancer

A nitrocellulose-coated antibody microarray featuring 84 antibodies was employed to screen for serum proteins that displayed significant difference for the lung cancer patients as compared to healthy controls and to chronic obstructive pulmonary disease (COPD) patients (Gao et al. 2005). Seven proteins including C-reactive protein, serum amyloid A, mucin 1, and alpha-1-antitrypsin showed relatively high specificity and sensitivity in identifying lung tumor patients from healthy subjects or patients with chronic pulmonary disease. Despite a poor correlation between mRNA and protein-expression levels, Han et al. (2009) designed a focused antibody microarray spotted with antibodies against 19 proteins selected based on their mRNA expression profiles and applied it to the profiling of proteins in 48 serum samples from lung cancer patients in an attempt to identify potential biomarkers of lung cancer. Six proteins were found to be distinctly downregulated in sera of lung cancer patients and in agreement with the mRNA profiling data. Blind test of 32 other human sera using the antibody microarray followed by hierarchical clustering analysis revealed an approximate sensitivity of 88%, specificity of 80%, and an accuracy of 84%, respectively, in classifying the sera, which supports the potential of the six identified proteins as biomarkers for the prognosis of lung cancer.

6.3.5 Bladder Cancer

Sanchez-Carbayo et al. (2006) designed an antibody microarray for bladder cancer by selecting antibodies against 183 proteins differentially expressed in bladder tumors based on the gene expression analysis results performed on bladder cancer biopsy samples. Protein profiles of serum samples obtained from the antibody microarrays distinguished bladder cancer patients from controls with a correct classification rate of 93.7%. Serum protein profiles derived from a second antibody microarray featuring a subset of the antibodies provided predictive information by stratifying patients with bladder tumors based on their overall survival. This study provides experimental evidence for the use of serum protein profiles obtained by antibody microarrays as comprehensive means for bladder cancer diagnosis and clinical outcome stratification.

6.3.6 Colorectal Cancer

Madoz-Gúrpide et al. (2007) used an early version of Panorama Ab Microarray-Cell Signaling from Sigma-Aldrich (St. Louis, MO) to examine alterations in cellular pathways in colorectal cancer (CRC) biopsy samples and identified a panel of 46 proteins that were differentially expressed between normal colorectal epithelium and adenocarcinoma. More than half of those proteins were further validated on a CRC tissue microarray and through immunoblotting analysis, which showed potential of being biomarkers for CRC. It was also found that a significant upregulation of the human epidermal growth factor receptor 1 (HER1/EGFR/ErbB1) and Wnt/beta-catenin pathways and the downregulation of p14 (ARF) were associated with CRC.

6.3.7 Esophageal Cancer

Uemura et al. (2009) examined surgical samples from three cases of esophageal cancer using the Panorama Antibody Microarray XPRESS Profiler725 (Sigma-Aldrich, St. Louis, MO) and identified 24 proteins with aberrant expression in comparison with the corresponding normal mucosa. Subsequent Western blotting and IHC validated six of these proteins. In addition to the identification of mitotic checkpoint proteins BubR1 and Mad2, which were previously reported to be involved in esophageal cancer, aberrant expressions of NFκB-activating kinase (NAK), caspase 10, and activator protein-1 (AP1) were observed in esophageal cancer, indicating that mitotic checkpoint proteins and NFκB play important roles in the carcinogenesis of esophageal cancer.

In addition, antibody microarrays have also been used for studies involved in characterizing the mechanisms of some anticancer drugs. An example of this approach is given by the work of Smith et al. (2006), in which they intended to identify biomarkers using the earlier version of the Panorama Cell Signaling Antibody Microarray featuring 224 antibodies; they were able to predict treatment responses of breast cancer patients to doxorubicin. Despite its effectiveness, doxorubicin resistance is a common occurrence among cancer patients. They compared the protein-expression profiles between the MDA-MB-231 breast cancer cell line and its doxorubicin-resistant derivative, and found that several proteins displaying differential expression patterns between the two cell lines were correlated with doxorubicin resistance, including ERKs, cyclins B1 and D2, cytokeratin 18, and heterogeneous nuclear ribonucleoprotein m3–m4. These results demonstrate that antibody microarrays can be used to identify novel biomarkers related to chemotherapy resistance, which in turn may allow therapy to be tailored to individual patients.

The early success of the antibody microarray technology in disease biomarker discovery has exemplified its clinical potential. While the translation of high-density antibody microarrays to the clinic has been very slow, recent years have witnessed moderate success in the areas of disease diagnostics and metabolite detection with increasing number of low-density antibody arrays or their variants available from companies involved in antibody array commercialization. OvaDx® is a microarray-based presymptomatic blood test for ovarian cancer developed by Arrayit Diagnostics (http://www.arrayit.com). It monitors the response of the immune system during early stages of ovarian tumor development through measuring about 100 biomarkers in serum samples.

As one of the leading players in transforming the antibody array technology for preclinical and clinical applications, Randox Laboratories (http://www.randox.com) has developed and marketed 14 antibody array-based multiplexed immunoassays for a wide range of markers present in serum or urine samples related to cardiac disease, metabolic syndrome, fertility, substance abuse, as well as cancer. Each of these arrays features from several up to a dozen antibodies in a format that allows 23 biological samples to be analyzed in parallel.

Developed for the use with its proprietary electrochemiluminescent detection system, several multiplex immunoassays have been developed on its Multi-Spot Plate surface by Meso Scale Discovery (MSD) (http://www.meso-scale.com), each of which targets up to three marker proteins associated with Alzheimer's disease, fertility, cardiac, and metabolic conditions in various bodily fluids. RayBiotech (http://www.raybiotech.com) carries a sandwich assay-based antibody array that is capable of detecting the expression of 182 adipokines in serum, indicators for human obesity. A group of antibody arrays based on the PATH® nitrocellulose slides, trademarked QPDx™ Biochips, are available from Avantra Biosciences (http://www. avantrabio.com) with various focuses such as angiogenesis, inflammation, sepsis, and adhesion.

Apart from the planar array format described above, Biosite Diagnostics Inc. (http://www.biosite.com) developed a series of antibody arrays for cardiovascular disease diagnosis and assessment, as well as for drug abuse screening, using its proprietary microcapillary fluidics technology and Triage® detecting system. Using Luminex microsphere-based immuno-multiplexing assay platform, Myriad-RBM (http://www.rulesbasedmedicine.com) has introduced a wide range of multianalyte profile (MAP) assays capable of tracking between 20 and 100 markers for cancer, cardiovascular diseases, kidney disorder, and schizophrenia. More significantly, Vermillion's OVA1 (http://www.vermillion.com), the first blood test cleared by the US Food and Drug Administration (FDA) for the evaluation of an ovarian adnexal mass prior to a planned surgery, was developed based on the five marker proteins uncovered with Ciphergen's SELDI-MS ProteinChip® technology.

Despite the promising leads illustrated by these preliminary data from antibody microarray-based studies using clinical samples, translation of the findings from these studies to the clinical uses has been hampered by the lack of independent validation approaches and the relatively small patient cohorts used in these studies. The latter can be attributed to the low sample throughput capability of the antibody microarray, which has been well compensated by another type of protein microarray, i.e., RPPMs.

6.4 Application of RPPMs in Disease Biomarker Validations and Drug Discovery

The concept of the RRPM technology was first introduced by Liotta's group back in 2000 and published in a seminal paper by Paweletz et al. (2001), based on a study examining several proliferation- and prosurvival pathway-related cell signaling proteins involved in prostate cancer progression using malignant tumor cells procured by LCM. Comparing samples derived from the epithelial cells from normal prostate, intraepithelial neoplasia, and invasive prostate cancer revealed a general increase in proteins involved in anti-apoptotic process in both premalignant and invasive samples, including protein kinases PKB/AKT and ERK1/2, indicating a possible correlation between the activation of prosurvival pathways and carcinogenesis in prostate (Paweletz et al. 2001).

An increasing number of RPPM-based studies in recent years have exemplified its broad application in mapping altered signal networks in individual clinical samples and demonstrated their potential utility in disease diagnosis and patient stratifications in the clinical settings with the goal of administrating patient-tailored therapy based on their respective abnormalities in underlying molecular pathways. So far, RPPMs have been applied in the studies of a wide range of human pathological conditions, including cancer, bacterial infections, and immunological disorders.

The diverse prognoses and therapeutic responses of cancer patients with similar pathologies indicate molecular heterogeneity of the disease, highlighting the importance and necessity of molecular profiling of cancer. As a result, one of the areas in which the RPPM approach has been actively explored is to identify altered proteins within signaling networks that may correlate with different histopathological types, grades, and stages of cancers. Due to the important roles they play in various aspects of cellular regulations, the expression of protein kinases and protein phosphatases along with their modification states, phosphorylation mainly, has been the focus of a number of RPPM-based studies. Wulfkuhle et al. (2003) examined the levels of phosphorylation on ERK1/2 [pT202 + pY204], PKB/AKT [pT308], PKB/AKT [pS473], and GSK3β [pY216] in ovarian cancers of different histotypes and stages. Not only was a high degree of variability found in the phosphorylation state of these protein kinases within each of the tumor types, but also no significant differences observed between different types or stages, indicating that the signaling pathways were altered in a patient-specific rather than histotype or stage-specific fashion, further emphasizing the importance of patient-tailored therapies. To understand the molecular mechanisms responsible for the poor prognosis of glioblastoma, the most advanced form of glioma of the central nervous system, Jiang et al. (2006) profiled the expression and phosphorylation state of 46 proteins involved in proliferation, survival, apoptosis, and invasion using a RPPM constructed with protein lysates derived from 37 glioblastoma and 45 lower-grade glioma tissues. A cluster of 12 proteins including IκBα, EGFR [pY845], PKB/AKT [pT308], and Bcl-2 displayed the most significant differences between the two groups, indicating their involvement in glioma progression and their potential as molecular markers for disease prognosis. Along the same vein, Petricoin et al. (2007) used RPPMs to dissect the abnormality of signaling pathways that might attribute to the lack of response to therapy in about one third of rhabdomyosarcoma and were able to establish a correlation between the activation of several effectors in the mTOR pathway and poor survival rate, which can be potentially used to select patients for appropriate therapeutic regimens.

Recently, advances in methodologies for extracting full-length proteins from FFPE tissues have further facilitated the implementation of RPPM technology in routine clinical settings (Berg et al. 2010, 2011). Daniela et al. (2012) analyzed signaling proteins involved in HER2-related pathways by RPPMs featuring lysates derived from 106 FFPE breast cancer tissue samples and found a significant correlation of HER2 with EGFR, human epidermal growth factor receptor 3 (HER3/ErbB3), and urokinase plasminogen receptor (uPAR) in routinely used FFPE breast cancer tissues, raising the potential of targeting EGFR, HER2, HER3, and uPAR together as an efficient treatment option for patients with breast cancer. A high concordance between FFPE breast cancer surgical specimens and their corresponding core biopsies in the expression levels of HER2, ERα, and PgR was also uncovered by the RPPM, affirming the value of core biopsy samples not only in breast cancer diagnosis but also for disease prognosis and therapeutic outcome prediction.

The advantage of the RPPM to work with limited amounts of clinical samples to generate readouts for multi-targets in a highly parallel fashion was demonstrated by a study by Tibes et al. (2006) using lysates derived from primary acute myelogenous leukemia (AML) cells and leukemic stem cells, the rare cell populations in clinical samples. The analysis of the RPPM spotted with lysates of CD34+/CD38+ and CD34+/CD38− cells from AML patients and from healthy bone marrow donors revealed increases in the levels of ERK1/2 [pT202 + Y204], PKB/AKT [pT308], and mTOR [pS2448] in AML stem cells (CD34+/CD38-) relative to leukemic or normal CD34+/CD38+ progenitor cells or normal CD34+/CD38− stem cells.

Frederick et al. (2011) used the RPPM to examine 60 phosphoproteins in matched tumor and nonmalignant biopsy specimens from 23 patients with head and neck squamous cell carcinoma. Among them, 18 were observed elevated in tumors versus healthy tissue whereas 17 displayed reductions. The most significant changes in tumor included checkpoint kinase 1 (CHK1) [pS345], CHK2 [pS33/S35], 4E-BP1 [pS65], PKC ζ/ι [pT410/T412], LKB1 [pS334], IκB-α [pS32], eIF4E [pS209], Smad2 [pS465/S467], IRS-1 [pS612], MEK1/2 [pS217/S221], and total PKC ι.

Nanjundan et al. (2010) compared 63 key signaling proteins involved in prosurvival, mitogenic, apoptotic, and growth pathways between paired non-small cell lung carcinoma (NSCLC) tumor and normal lung specimens using the RPPM. Expression of a cluster of 15 proteins was found significantly different between normal lung and NSCLC tumors. A signature panel of proteins including caveolin-1, p70S6K, cyclin B1, and Src [pY527] was identified to be able to distinguish tumors from normal lung, which was subsequently validated by immunohistochemistry.

Currently, similar therapies are applied to treat both primary and metastatic tumors based on the notion that they share similar molecular defects and should response similarly to therapy. Recently, the Wulfkuhle's group compared the phosphorylation states of 128 key signaling proteins in brain metastases developed from patients with breast or NSCLC using the epithelial cell populations procured by LCM on RPPMs. The activation of the HER2/IGFR-AKT pathways was observed in breast cancer brain metastases whereas increased levels of proteins involved in the EGFR-ERK signaling pathway were detected in NSCLC metastases, indicating an organ-specific alteration in molecular pathways in cancer metastases, which underscored the importance of designing patient-tailored therapeutic regimens in treating metastatic cancer patients (Improta et al. 2011).

Apart from the use with biopsy/tissue samples, the RPPM was also shown as a viable approach for the detection of biomarkers in serum samples. Grote et al. (2008) examined the level of a pancreatic serum marker protein, CA19.9, on the RPPM printed with serum and plasma samples from pancreatic cancer patients. The study showed a strong data correlation between RPPM and ELISA with a higher sensitivity observed in the former, which supported RPPM as a promising technique for rapid screening and validation of candidate biomarkers in serum. Gyorgy et al. (2010) examined the levels of neuronal and glial proteins including S100B, neuron-specific enolase (NSE), myelin basic protein (MBP), and neurofilament heavy chain (NF-H) in the peripheral circulating blood over time after traumatic brain injury (TBI) using the RPPM technique. Although serum levels of S100B, MBP, and NF-H showed a time-dependent increase following injury, only serum NF-H levels seemed to correlate with poor clinical outcome, indicating that it might be a useful diagnostic indicator for injury severity.

The potential of the RPPM has also been explored in the early detection of cancer, on the basis of the notion that cancer patients spontaneously produce autoantibodies toward tumor-derived proteins. A custom protein microarray featuring about 5,000 candidate tumor antigens was applied to screen autoantibodies in the sera from patients with early stage breast cancer. A panel of 28 proteins was identified as a classifier with a sensitivity of 80.8% and a specificity of 61.6% for the early detection of breast cancer (Anderson et al. 2011).

Compared to antibody microarrays, the RPPM technology has advanced much more rapidly toward clinical applications, most likely due to the higher level of technical maturity and higher data accuracy. The RPPM has been utilized in several clinical trials for molecular profiling of the activation state of pathways targeted by a drug as well as screening for off-target effects, aiming at providing insights into the molecular action of the drug and the underlying mechanisms of differences in pathway activation that might contribute to the differential responses to the drug by the same group of patients. The RPPM was typically applied in the clinical trials as part of integrated technology platform, which usually includes medical imaging, comparative genomic hybridization (CGH), and gene expression profiling (Mueller et al. 2010).

A recent search of the clinical trial information database, clinical trials (http://www.clinicaltrials.gov), at the US National Institutes of Health turned up several studies involving RPPMs that can be broadly categorized into two directions. Among them, the RPPM analysis has been used as a molecular profiling tool for disease biomarker identifications for disease classification and patient stratification. One of the ongoing trials is to analyze global histone protein and non-histone expression and covalent modification profiles from lymphoid and stromal cells across a broad panel of lymphomas using reverse-phase protein microarrays with a panel of protein-specific and/or modification-specific antibodies (NCT00952809). The RPPM analysis is also being employed in a pilot study as part of the molecular profiling strategy including IHC, FISH, DNA microarray for the identification of potential targets and selection of treatment regimes for patients with metastatic breast cancer (NCT01074814).

Three ongoing clinical trials of several compounds and biologics targeting various human solid tumors were documented using the RPPM analysis as one of the approaches used to monitor patient responses and treatment outcome. A phase Ib study of MK-2206, an AKT/PKB inhibitor, in combination with standard doses of paclitaxel, is currently under way in patients with metastatic breast cancer or other metastatic solid tumors to determine its side effects and best dose of MK2206 when given with paclitaxel. In this study, the RPPA is one of the secondary measures for monitoring the changes in molecular profile before and after treatment (NCT01263145). The RPPM is also involved in a phase II study of irinotecan with panitumumab as second-line therapy for patients with advanced esophageal adenocarcinoma, in which it measures the expression of the EGF-pathway associated proteins and their phosphorylation states and correlates the data with patient response after the treatment (NCT00836277). The technology is being used to evaluate efficacy of the combined therapy of postoperative chemoradiotherapy and panitumumab in patients with advanced head and neck cancer (NCT00798655).

In addition to cancers, the RPPM has also been applied to the profiling of signaling pathway in other human diseases. To understand how defects in natural killer cells result in type 1 diabetes in children, Qin and colleagues monitored the activation state of NKG2D-mediated phosphoinositide 3-kinase–AKT pathway and showed clear defects in the activation of PI3K and its downstream effector AKT/PKB, which is in agreement with an observed reduction in NKG2D-dependent cytotoxicity and interferon-γ secretion (Qin et al. 2011).

Recently, the RPPM approach was developed and validated as an assay platform for cell-based screening of protein kinase inhibitors (van Oostrum et al. 2009). The RPPMs spotted with lysates of A431 cells treated with a panel of 12 kinase inhibitors of various protein kinases including EGFR, ErbB2, IGF1R, FAK, PI3K, and MEKs individually were probed with select pre-validated antibodies to characterize the effects of these inhibitors on the insulin- and EGF-mediated signaling pathways. The study demonstrated the suitability of the RPPM technology for cell-based screening of compounds acting as kinase inhibitors.

The RPPM technology was also recently employed in a high-throughput screening for microRNAs that regulate estrogen receptor (ER) signaling in breast cancer cells (Leivonen et al. 2009). Systematic monitoring of the ER protein levels after high-throughput transfections of MCF-7 and BT-474 cells with a pre-miR miRNA library containing 319 pre-miR constructs revealed 21 miRNAs that downregulated the ERα, which were validated by Western blotting and quantitative real-time PCR, and by demonstrating the inhibition of estrogen-stimulated cell growth.

The RPPM assay format has been successfully incorporated into the technology platforms of several molecular diagnostic companies aimed at developing novel diagnostics for severe chronic diseases and the associated treatment based on autoantibody signatures in serum or plasma. QuantiSpot Rheumatoid Arthritis (RA) Assay RA Panel from SQI Diagnostics (http://www.sqidiagnostics.com) has recently been cleared by US FDA for simultaneously measuring rheumatoid factor (RF) (IgA, IgG, and IgM) and cyclic citrullinated peptide (CCP) IgG, providing physicians the tools necessary to detect RA patients of early stage sooner and predict disease progression faster. Protagen AG (http://www.protagen.de) takes advantage of its proprietary UNIarray® platform containing the largest collection of recombinant human proteins from five tissue-specific recombinant human protein-expression libraries for the discovery of novel biomarkers based on indication-specific autoantibody signatures in autoimmune diseases, cancer, and neurodegenerative disorders. ProtoArrays® Pathogen Microarrays from Life technologies (http://www.invitrogen.com) permits parallel detection of infectious agents in blood samples including *B. anthracis*, vaccinia, monkeypox, *F. tularensis* (tularemia), *Y. pestis* (plague), Ebola, Marburg, and dengue virus 1, 2, 3, and 4.

6.5 Challenges for the Clinical Application of Protein Microarrays

Increasing recognition of the limitations of genomic information in providing insights into physiological and pathological processes in combination with technological advancements in array technologies in the past decade has provoked strong interest in protein microarrays. Early success in the preliminary studies on various human diseases using protein microarrays has made the clinical application of the technology well within reach.

However, despite strong interests and intensive efforts from both the academic and the clinical communities, the promises demonstrated by early studies have not been delivered so far. Even though originated around the same time and based on the same basic principle of microimmunoassays, antibody microarrays and RPPMs have been progressing at very different paces during their integration process into routine clinical practice, probably due to the underlying differences in assay configurations and the resulting data quality.

The antibody microarray was developed as a research tool for large-scale profiling of hundreds and even thousands of proteins in a parallel fashion within a single sample. Perhaps, the greatest barrier to translation of putative biomarkers discovered in the protein microarray-based studies is the inability to evaluate them in high-throughput validation with large patient cohorts. Challenges for the widespread use of antibody microarrays in routine clinical settings include the low sample throughput capability and the concerns over data quality, accuracy, and reproducibility from antibody microarrays. Therefore, it is always imperative to confirm resulting data using independent validation approaches. A growing number of panels of proteins with the potential of being disease biomarkers have been identified over the years for a wide range of human diseases through the studies where antibody microarray technology was involved. However, none of them have been adopted and cleared for use in the clinic so far, highlighting the challenges involved in translating of discoveries made on the bench to use at the bedside.

Much of the effort in the field of antibody microarrays has been focused on the development of antibody microarrays with broader target coverage. While antibody microarrays with the ability to track a wide range of proteins may offer a greater advantage as a discovery tool in basic research, fulfilling their long-standing promise as a practical tool for disease screening, diagnostics, prognosis, and therapeutic response prediction calls for nothing more than focused arrays spotted with a handful of highly validated antibodies targeting a panel of well-characterized biomarkers discovered through initial large-scale screenings. Printing such focused microarrays into microtiter plate wells will not only increase the sample throughput capability allowing hundreds of clinical samples to be analyzed simultaneously but will also enable seamless integration of microarray technology with standard clinical laboratory instruments for automated sample handling and analysis.

In addition, direct or indirect labeling of proteins in clinical samples for signal detection is tedious and impractical in the clinical settings, not to mention the possibility that the labeling may mask the antigenic epitopes, resulting in physical interference with the ability of the antibody to interact with its target protein. Furthermore, the commonly used fluorescence dye-based detection is not compatible for monitoring some analytes due to their inherent autofluorescence property, which decreases the signal-to-noise ratio. Thus, developing antibody microarray assays with label-free detection is vital for the implementation of technology in the clinic, not only to avoid any undesired effects as a result of protein modifications but also for ease of clinical sample processing. As a result, a number of label-free detection methods have been developed, including surface plasmon resonance (SPR), semiconducting carbon nanotubes, and microcantilever sensors (Ramachandran et al. 2006; Chandra et al. 2011), which have shown some promising results. This is a vital issue that needs to be addressed prior to the full integration of the antibody microarray analysis into the clinic.

In contrast, the RPPM has been making huge strides moving toward clinical applications and has been integrated into several ongoing clinical trials either for the identification of therapeutic targets or to monitor biological effects of drug candidates (pharmacodynamic biomarkers) since 2005. The ability of the RPPM to work with minute amounts of samples in a high-throughput fashion and to have a detection limit in the picomolar concentration range where many of proteins of clinical importance are found might have led to the early delivery of RPPM in clinical settings. One of main hurdles that had hindered the clinical application of RPPM was the impossibility of extracting immunoreactive protein molecules from routine clinical FFPE samples, which was recently overcome after several FFPE protein extraction protocols were developed (Berg et al. 2010, 2011; Daniela et al. 2012).

An additional challenge for the implementation of the RPPM in clinical settings is the reproducibility and comparability of data from different arrays analyzed at different locations. Therefore, it is very important to standardize the protocols for every step involved in the RPPA analysis, including sample procurement, processing, antibody characterization, signal detection, and data processing and analysis. The inclusion of common, renewable reference samples such as a pooled cell or tissue lysate, an antigenic peptide cocktail, or a housekeeping protein across all the RPPAs as the internal control will allow better data normalization and is critical for comparing data generated on different arrays.

6.6 Future Perspectives

The field of protein microarrays has come a long way since just over a decade ago when the very first "homemade" macroarray with only a few dozens of antibodies spotted by hand was reported. Today, through multidisciplinary collaborative efforts, protein microarrays featuring hundreds even thousands of proteins or antibodies are readily available from various commercial sources either as stand-alone products or as part of integrated services. With the expansion of the repertoire of recombinant proteins and better-characterized antibodies or antibody-like molecular probes, the cost per data point has dropped significantly, facilitating the widespread applications of protein microarrays in both basic research and the clinical settings.

The potential clinical utility of protein microarrays in disease diagnostics and therapeutics is evidenced by the results from many of exploratory studies involving clinical samples. On the one hand, the antibody microarray holds great potential in disease biomarker screening. On the other hand, the RPPM has proved its utility in molecular profiling of signaling pathways in disease states to provide information aiding patient stratification, therapeutic regimen selection, and treatment efficacy prediction. Furthermore, their integration along with combining the power of mass spectrometry-based de novo proteomic technology with the high sensitivity of protein microarrays should have a synergistic effect on disease biomarker research, paving the way for the development of a powerful and comprehensive biomarker discovery platform. However, the realization of the full potential of protein microarrays in the clinical applications awaits continued optimization and standardization of the technologies associated with sample collection and preparation, microarray substrate production, array printing, sample preparation, and signal detection, as well as better study designs. Thus, implementation of protein microarray assays as part of clinical routine practice may still be several years away.

It is expected that much of the effort in the next few years will be focused on the creation of a proteome-wide protein microarray capable of tracking most of the proteins coded by the estimated 23,000 human genes in the human genome along with their various posttranslational modification states. On a separate front, the focus of the protein microarray field will be shifted beyond the initial discovery stage of biomarker candidates toward verifications and validations (Rifai et al. 2006). The use of larger, carefully selected patient cohorts in protein microarray-based studies and the application of rigorous, orthogonal validation approaches are the two critical needs to be met before the translation of any biomarker candidates to the clinic.

References

Anderson KS, Sibani S, Wallstrom G, Qiu J, Mendoza EA, Raphael J, Hainsworth E, Montor WR, Wong J, Park JG, Lokko N, Logvinenko T, Ramachandran N, Godwin AK, Marks J, Engstrom P, Labaer J (2011) Protein microarray signature of autoantibody biomarkers for the early detection of breast cancer. J Proteome Res 10:85–96

Angenendt P, Glökler J, Sobek J, Lehrach H, Cahill DJ (2003) Next generation of protein microarray support materials: evaluation for protein and antibody microarray applications. J Chromatogr A 1009:97–104

Baehner FL, Lee M, Demeure MJ, Bussey KJ, Kiefer JA, Barrett MT (2011) Genomic signatures of cancer: basis for individualized risk assessment, selective staging and therapy. J Surg Oncol 103:563–573. doi:10.1002/jso.21838

Baldrich E (2011) Aptamer arrays. Methods Mol Biol 671:35–54

Berg D, Hipp S, Malinowsky K, Böllner C, Becker KF (2010) Molecular profiling of signalling pathways in formalin-fixed and paraffin-embedded cancer tissues. Eur J Cancer 46:47–55

Berg D, Langer R, Tran K, Walch A, Schuster T, Bronger H, Becker KF (2011) Protein microarray-based comparison of HER2, estrogen receptor, and progesterone receptor status in core biopsies and surgical specimens from FFPE breast cancer tissues. Appl Immunohistochem Mol Morphol 19:300–305

Carlsson A, Wingren C, Ingvarsson J, Ellmark P, Baldertorp B, Fernö M, Olsson H, Borrebaeck CA (2008) Serum proteome profiling of metastatic breast cancer using recombinant antibody microarrays. Eur J Cancer 44:472–480

Chandra H, Reddy PJ, Srivastava S (2011) Protein microarrays and novel detection platforms. Expert Rev Proteomics 8:61–79

Daniela B, Claudia W, Katharina M, Kai T, Axel W, Holger B, Holger B, Tibor S, Heinz H, Karl-Friedrich B (2012) Profiling signalling pathways in formalin-fixed and paraffin-embedded breast cancer tissues reveals cross-talk between EGFR, HER2, HER3 and uPAR. J Cell Physiol 227:204–212. doi:10.1002/jcp.22718

Davies AH, Barrett I, Pambid MR, Hu K, Stratford AL, Freeman S, Berquin IM, Pelech S, Hieter P, Maxwell C, Dunn SE (2011) YB-1 evokes susceptibility to cancer through cytokinesis failure, mitotic dysfunction and HER2 amplification. Oncogene 30:3649–3660. doi:10.1038/onc.2011.82

Ekins RP, Chu FW (1991) Multianalyte microspot immunoassay–microanalytical "compact disk" of the future. Clin Chem 37:1955–1967

Frederick MJ, VanMeter AJ, Gadhikar MA, Henderson YC, Yao H, Pickering CC, Williams MD, El-Naggar AK, Sandulache V, Tarco E, Myers JN, Clayman GL, Liotta LA, Petricoin EF 3, Calvert VS, Fodale V, Wang J, Weber RS (2011) Phosphoproteomic analysis of signaling pathways in head and neck squamous cell carcinoma patient samples. Am J Pathol 178:548–571

Fujita K, Ewing CM, Sokoll LJ, Elliott DJ, Cunningham M, De Marzo AM, Isaacs WB, Pavlovich CP (2008) Cytokine profiling of prostatic fluid from cancerous prostate glands identifies cytokines associated with extent of tumor and inflammation. Prostate 68:872–882

Gao WM, Kuick R, Orchekowski RP, Misek DE, Qiu J, Greenberg AK, Rom WN, Brenner DE, Omenn GS, Haab BB, Hanash SM (2005) Distinctive serum protein profiles involving abundant proteins in lung cancer patients based upon antibody microarray analysis. BMC Cancer 5:110

Geierstanger BH, Saviranta P, Brinker A (2006) Antibody microarrays using resonance light-scattering particles for detection. Methods Mol Biol 328:31–50

Grote T, Siwak DR, Fritsche HA, Joy C, Mills GB, Simeone D, Whitcomb DC, Logsdon CD (2008) Validation of reverse phase protein array for practical screening of potential biomarkers in serum and plasma: accurate detection of CA19-9 levels in pancreatic cancer. Proteomics 8:3051–3060

Gyorgy AB, Walker J, Wingo D, Eidelman O, Pollard HB, Molnar A, Agoston DV (2010) Reverse phase protein microarray technology in traumatic brain injury. J Neurosci Methods 192:96–101

Haab BB, Dunham MJ, Brown PO (2001) Protein microarrays for highly parallel detection and quantitation of specific proteins and antibodies in complex solutions. Genome Biol 2:1–22

Han MK, Oh YH, Kang J, Kim YP, Seo S, Kim J, Park K, Kim HS (2009) Protein profiling in human sera for identification of potential lung cancer biomarkers using antibody microarray. Proteomics 9:5544–5552

Hanahan D, Weinberg RA (2011) Hallmarks of cancer: the next generation. Cell 144:646–674

Hudelist G, Pacher-Zavisin M, Singer CF, Holper T, Kubista E, Schreiber M, Manavi M, Bilban M, Czerwenka K (2004) Use of high-throughput protein array for profiling of differentially expressed proteins in normal and malignant breast tissue. Breast Cancer Res Treat 86:281–291

Improta G, Zupa A, Fillmore H, Deng J, Aieta M, Musto P, Liotta LA, Broaddus W, Petricoin EF, Wulfkuhle JD (2011) Protein pathway activation mapping of brain metastasis from lung and breast cancers reveals organ type specific drug target activation. J Proteome Res 10:3089–3097, PubMed PMID:21574647

Ingvarsson J, Wingren C, Carlsson A, Ellmark P, Wahren B, Engström G, Harmenberg U, Krogh M, Peterson C, Borrebaeck CA (2008) Detection of pancreatic cancer using antibody microarray-based serum protein profiling. Proteomics 8:2211–2219

Jiang R, Mircean C, Shmulevich I, Cogdell D, Jia Y, Tabus I, Aldape K, Sawaya R, Bruner JM, Fuller GN, Zhang W (2006) Pathway alterations during glioma progression revealed by reverse phase protein lysate arrays. Proteomics 6:2964–2971

Knezevic V, Leethanakul C, Bichsel VE, Worth JM, Prabhu VV, Gutkind JS, Liotta LA, Munson PJ, Petricoin EF 3, Krizman DB (2001) Proteomic profiling of the cancer microenvironment by antibody arrays. Proteomics 1:1271–1278

Koga H, Kyo M, Usui-Aoki K, Inamori K (2006) A chip-based miniaturized format for protein-expression profiling: the exploitation of comprehensively produced antibodies. Electrophoresis 27:3676–3683

Leivonen SK, Mäkelä R, Ostling P, Kohonen P, Haapa-Paananen S, Kleivi K, Enerly E, Aakula A, Hellström K, Sahlberg N, Kristensen VN, Børresen-Dale AL, Saviranta P, Perälä M, Kallioniemi O (2009) Protein lysate microarray analysis to identify microRNAs regulating estrogen receptor signaling in breast cancer cell lines. Oncogene 28:3926–3936

MacBeath G (2002) Protein microarrays and proteomics. Nat Genet 32:526–532

Madoz-Gúrpide J, Cañamero M, Sanchez L, Solano J, Alfonso P, Casal JI (2007) A proteomics analysis of cell signaling alterations in colorectal cancer. Mol Cell Proteomics 6:2150–2164

Mannsperger HA, Uhlmann S, Schmidt C, Wiemann S, Sahin O, Korf U (2010) RNAi-based validation of antibodies for reverse phase protein arrays. Proteome Sci 8:69

Miller JC, Zhou H, Kwekel J, Cavallo R, Burke J, Butler EB, Teh BS, Haab BB (2003) Antibody microarray profiling of human prostate cancer sera: antibody screening and identification of potential biomarkers. Proteomics 3:56–63

Mueller C, Liotta LA, Espina V (2010) Reverse phase protein microarrays advance to use in clinical trials. Mol Oncol 4:461–481

Nanjundan M, Byers LA, Carey MS, Siwak DR, Raso MG, Diao L, Wang J, Coombes KR, Roth JA, Mills GB, Wistuba II, Minna JD, Heymach JV (2010) Proteomic profiling identifies pathways dysregulated in non-small cell lung cancer and an inverse association of AMPK and adhesion pathways with recurrence. J Thorac Oncol 5:1894–1904

Orchekowski R, Hamelinck D, Li L, Gliwa E, vanBrocklin M, Marrero JA, Vande Woude GF, Feng Z, Brand R, Haab BB (2005) Antibody microarray profiling reveals individual and combined serum proteins associated with pancreatic cancer. Cancer Res 65:11193–11202

Paweletz CP, Charboneau L, Bichsel VE, Simone NL, Chen T, Gillespie JW, Emmert-Buck MR, Roth MJ, Petricoin EF III, Liotta LA (2001) Reverse phase protein microarrays which capture disease progression show activation of pro-survival pathways at the cancer invasion front. Oncogene 20:1981–1989

Petricoin EF 3, Espina V, Araujo RP, Midura B, Yeung C, Wan X, Eichler GS, Johann DJ Jr, Qualman S, Tsokos M, Krishnan K, Helman LJ, Liotta LA (2007) Phosphoprotein pathway mapping: akt/mammalian target of rapamycin activation is negatively associated with childhood rhabdomyosarcoma survival. Cancer Res 67:3431–3440

Qin H, Lee IF, Panagiotopoulos C, Wang X, Chu AD, Utz PJ, Priatel JJ, Tan R (2011) Natural killer cells from children with type 1 diabetes have defects in NKG2D-dependent function and signaling. Diabetes 60:857–866

Ramachandran N, Larson DN, Stark PR, Hainsworth E, LaBaer J (2006) Emerging tools for real-time label-free detection of interactions on functional protein microarrays. FEBS J 272:5412–5425

Rifai N, Gillette MA, Carr SA (2006) Protein biomarker discovery and validation: the long and uncertain path to clinical utility. Nat Biotechnol 24:971–983

Sanchez-Carbayo M, Socci ND, Lozano JJ, Haab BB, Cordon-Cardo C (2006) Profiling bladder cancer using targeted antibody arrays. Am J Pathol 168:93–103

Schweitzer B, Roberts S, Grimwade B, Shao W, Wang M, Fu Q, Shu Q, Laroche I, Zhou Z, Tchernev VT, Christiansen J, Velleca M, Kingsmore SF (2002) Multiplexed protein profiling on microarrays by rolling-circle amplification. Nat Biotechnol 20:359–365

Sevecka M, Wolf-Yadlin A, MacBeath G (2011) Lysate microarrays enable high-throughput, quantitative investigations of cellular signaling. Mol Cell Proteomics 10(4):M110.005363

Shao W, Zhou Z, Laroche I, Lu H, Zong Q, Patel DD, Kingsmore S, Piccoli SP (2003) Optimization of rolling-circle amplified protein microarrays for multiplexed protein profiling. J Biomed Biotechnol 2003:299–307

Silvestri A, Colombatti A, Calvert VS, Deng J, Mammano E, Belluco C, De Marchi F, Nitti D, Liotta LA, Petricoin EF, Pierobon M (2010) Protein pathway biomarker analysis of human cancer reveals requirement for upfront cellular-enrichment processing. Lab Invest 90:787–796

Smith L, Watson MB, O'Kane SL, Drew PJ, Lind MJ, Cawkwell L (2006) The analysis of doxorubicin resistance in human breast cancer cells using antibody microarrays. Mol Cancer Ther 5:2115–2120

Tibes R, Qiu Y, Lu Y, Hennessy B, Andreeff M, Mills GB, Kornblau SM (2006) Reverse phase protein array: validation of a novel proteomic technology and utility for analysis of primary leukemia specimens and hematopoietic stem cells. Mol Cancer Ther 5:2512–2521

Uemura N, Nakanishi Y, Kato H, Nagino M, Hirohashi S, Kondo T (2009) Antibody-based proteomics for esophageal cancer: Identification of proteins in the nuclear factor-kappaB pathway and mitotic checkpoint. Cancer Sci 100:1612–1622

van Oostrum J, Calonder C, Rechsteiner D, Ehrat M, Mestan J, Fabbro D, Voshol H (2009) Tracing pathway activities with kinase inhibitors and reverse phase protein arrays. Proteomics Clin Appl 3:412–422

Wulfkuhle JD, Aquino JA, Calvert VS, Fishman DA, Coukos G, Liotta LA, Petricoin EF 3 (2003) Signal pathway profiling of ovarian cancer from human tissue specimens using reverse-phase protein microarrays. Proteomics 3:2085–2090

Chapter 7
Peptide Arrays

Hong Zhang, Steven Pelech, Rob Ruijtenbeek, Thomas Felgenhauer, Ralf Bischoff, Frank Breitling, and Volker Stadler

7.1 General Presentation of Peptide Arrays with Some Examples of Implementation

Hong Zhang and Steven Pelech

7.1.1 Introduction

The completion of the human genome sequencing project has led to a major shift in biomedical research from pure genomics to functional genomics and proteomics, aiming at better understanding of the function of the genes or more appropriately their protein products. Although the DNA/oligonucleotide microarray technology

H. Zhang • S. Pelech (✉); Corresponding author for Section 7.1
Kinexus Bioinformatics Corporation, Vancouver, BC, Canada
e-mail: spelech@kinexus.ca

R. Ruijtenbeek (✉); Corresponding author for Section 7.2
PamGene International, B.V. s'Hertogenbosch, The Netherlands
e-mail: rruijtenbeek@pamgene.com

T. Felgenhauer (✉); Corresponding author for Section 7.3 • V. Stadler
PEPperPRINT GmbH, Rischerstrasse 12, 69123 Heidelberg, Germany
e-mail: thomas.felgenhauer@pepperprint.com

R. Bischoff
Research Group Chip-based peptide libraries, German Cancer Research Center, Im Neuenheimer Feld 580, 69120 Heidelberg, Germany

F. Breitling
Karlsruhe Institute of Technology, Hermann-von-Helmholtz-Platz 1, 76344 Eggenstein-Leopoldshafen, Germany

B. Jordan (ed.), *Microarrays in Diagnostics and Biomarker Development*,
DOI 10.1007/978-3-642-28203-4_7,

has proved to be unparalleled in its power in profiling gene expression and identifying single nucleotide polymorphisms (SNPs) in high-throughput manner, there is a poor correlation between mRNA and protein expression and activity. This arises from regulation of mRNA translation, for example, with microRNAs, and from the involvement of posttranscriptional and posttranslational modifications (PTMs), protein–protein interactions, and differences in subcellular localizations. These confounding factors have driven the development and the use of array technologies to directly study proteomes and have set the stage for the arrival of microarray-based proteomics.

Similarly to DNA/oligonucleotide microarrays, arrays for proteomics studies feature a wide range of molecules including recombinant proteins, complex protein samples, antibodies, peptides, or small molecules that are assembled in an addressable fashion on planar surfaces to allow parallel interrogations for activity and interactions associated with biomolecules at the protein level. Among various proteomics array types, protein microarrays and antibody microarrays have attracted most of attention in the field in the past decade, as illustrated by the rapid technological advancements and broad applications in numerous basic research and clinical studies (reviewed in Chap. 6).

Even though synthetic peptides together with oligonucleotides were among the first to be explored as molecular probes in array format, the development and application of peptide arrays has been sluggish and has lagged far behind arrays of many other types (Frank et al. 1983; Geysen et al. 1984). One of the concerns with the use of peptides as opposed to proteins as molecular probes is the potential loss of information as a result of the missing structural context (Mahrenholz et al. 2010). There is a much higher degree of entropy in the structures of peptides than with macromolecules such as proteins which are much more constrained. Arrays of proteins have the distinct advantage in being able to mimic their physiological counterparts through presenting full-length proteins that are likely in their native 3D conformations and with proper PTMs and associated proteins, to facilitate investigation of protein–protein interactions and assay of physiological activities. However, the molecular cloning and expression of tens of thousands of protein-coding genes with optimal covalent modifications and in combination with activating subunits has been proven to be technologically challenging, which significantly hampers the development and application of protein microarrays.

In contrast, the chemistry of solid-phase peptide synthesis (SPPS) was well defined about half a century ago (Merrifield 1965a, b). Further technological advancement in peptide synthesis over the years coupled with the recent influx of genome sequencing information upon the completion of a number of genome-wide sequencing projects has made designing and synthesizing peptides with defined sequences corresponding to the segments of a protein easier than ever. Furthermore, given the fact that the biological activity of a protein is often carried out through the coordinated actions of individual protein domains, it is reasonable to expect that the various functions of the protein can be recapitulated through the study of its constituent peptides representing individual functional domain sequences.

Since its advent about two decades ago, the peptide array has been implemented in a broad range of applications including antibody screening and epitope mapping, characterization of molecular interactions, and enzymatic activity profiling and has since become an increasingly important and versatile tool for proteomics research. More recently, with the improvements in array production techniques, peptide microarrays with higher peptide density and diversity can be produced en masse using the peptides generated from the parallel synthesis approaches such as the SPOT technology. The utility of these peptide microarrays has been demonstrated in some large-scale systems biology studies on the dynamics of protein interactions in cell signaling networks as well as kineome (also known as kinome) activity profiling. However, the clinical application of peptide arrays is still at its infancy despite potential and promises shown in early exploratory work.

This chapter summarizes the development of peptide array technology and highlights the progress made toward its applications in proteomics research in recent years. A perspective of its future implementation in clinical practice is also presented.

7.1.2 Development of Peptide Arrays

Despite the fact that the concept of the array-format peptide synthesis was introduced about three decades ago, it was not until the introduction of the SPOT synthesis of peptides in the early 1990s that the peptide array finally took its shape (Geysen et al. 1984; Frank 1992). Currently, there are two main strategies (in addition to very new approaches, see Sect. 7.2) being used to produce peptide arrays: in situ parallel on-chip synthesis or immobilization of presynthesized peptides on the array surface. Each has its own advantages and shortcomings. Hence, the choice of the approach for peptide array production is largely determined by the downstream applications of the resulting arrays.

7.1.2.1 Peptide Arrays by In situ Synthesis

The in situ on-chip synthesis of peptides is affordable as a result of the requirement of small amounts of reagents and of the fact that no purification of individual peptides is required. However, by the same token, the resulting peptides from in situ synthesis may suffer from low purity due to the variability in coupling efficiency between amino acid residues, especially in the case of long peptides (>20 residues) or those with residues such as Cys or Met, or containing multiple hydrophobic residues or phosphorylated amino acid residues. The two most common techniques for in situ synthesis that have been described and routinely used are SPOT synthesis and photolithography.

The SPOT synthesis was introduced by Ronald Frank back in 1992 and is essentially a stepwise synthesis of peptides through sequentially delivering small amounts of activated amino acids on functionalized cellulose or polypropylene membranes using standard Fmoc-based peptide chemistry (Frank 1992). The resulting membrane-based arrays are usually at low to medium density and can be used directly for downstream assays such as antibody epitope mapping. Recently, improvements have been introduced allowing peptides either to be cleaved from the membrane using strong bases or to be recovered from individual spots as soluble peptides. For this purpose, they are synthesized on acid-labile cellulose membranes (e.g., trifluoroacetic acid (TFA)-soluble), allowing postsynthesis printing of the recovered peptides on selected array surfaces at a higher density (Zander et al. 2005; Hilpert et al. 2007).

Another technique used in preparing peptide arrays in situ is the photolithographic synthesis developed by Fodor et al. (1991a, b) on the basis of the addressable surface activation concept. Compared to SPOT synthesis, photolithographic synthesis of peptides on array surface is more suitable for generating high-density arrays. However, it is both laborious and expensive. It requires the use of special photolabile protected amino acid derivatives as building blocks and of photomasks through which a laser is then used to activate specific areas on the array to cleave photolabile protecting groups. Even though the technique was initially developed for peptide synthesis, it was more easily adopted for the production of oligonucleotide arrays, due to the complexity of making masks for each of the 20 amino acids for every coupling cycle, as opposed to only four bases in oligonucleotide array production (Pease et al. 1994; McGall and Fidanza 2001). In recent years, a number of modifications have been introduced, including the uses of conventional amino acids and photogenerated reagents, resulting in the improvement of efficiency and the reduction in cost for array production (Singh-Gasson et al. 1999; Gao et al. 2003; Bhushan 2006; Shin et al. 2010).

7.1.2.2 Peptide Arrays by Postsynthesis Printing

Thanks to technological advances in microarray printing and array substrate production in the field of genomics, it has become a common practice to spot presynthesized peptides onto reactive planar array surfaces. The approach is particularly useful when multiple copies of the same array with high density are required. Furthermore, peptides can be purified after synthesis and prior to array printing to avoid any complications that may stem from peptide impurity. Various chemistries have been utilized for immobilizing peptides onto the array surfaces. One of the popular approaches is to attach N-terminal biotinylated peptides onto avidin/streptavidin-coated microarray slides (Lesaicherre et al. 2002). Covalently immobilizing peptides through a terminal Cys onto a surface functionalized with maleimide groups or disulfide is also a method that is being used quite commonly (Inamori et al. 2008). The suitability of immobilization chemistry varies according to the nature of the peptide as well as the downstream applications of the array. In our hands, attaching peptides through the N-terminal free amino groups to a surface functionalized with epoxide groups worked very well for assaying protein kinase activity (see below). Thus, it is very important and necessary to determine which strategy should be used for preparing peptide arrays already prior to peptide synthesis, according to the intended application of the resulting arrays.

7.1.3 Application of Peptide Arrays in Proteomics Studies

Since the advent of the technology more than two decades ago, peptide arrays have been applied in a broad range of investigations including antibody epitope mapping, protein domain-mediated interaction screening, and enzymatic activity profiling. In recent years, the utility of peptide arrays has been further extended to system-wide proteomics studies, fuelled by the advances in genomics and proteomics. This can be exemplified by their roles in cell signaling studies.

7.1.3.1 Enzymatic Activity Assays

Kineome Activity Profiling

Reversible phosphorylation and dephosphorylation of proteins mediated by protein kinases and protein phosphatases, respectively, are recognized as one of the most important and widespread molecular mechanisms in regulating cell signaling pathways involved in cell proliferation, division, differentiation, adherence, angiogenesis, and apoptosis (Brognard and Hunter 2011). According to our most recent tallies, the total number of phosphorylation sites in the human proteome is estimated to exceed 650,000, which encompass over 100,000 phosphosites that have been experimentally characterized and those predicted based on evolutionary conservation (http://www. phosphoNET.ca). The biological importance and potential clinical significance of protein phosphorylation, exemplified by the implication of deregulation of both kinase and phosphatase activity in a wide range of human diseases including cancer, autoimmune diseases, neurodegenerative diseases, and diabetes, has driven the development of strategies for the identification of physiological substrates for each of over 500 protein kinases within the human kineome, as well as for systematic profiling of kinase activity in biological samples.

For the identification of physiological substrates of kinases, a protein microarray featuring all the proteins representing the entire proteome would seemingly be an ideal platform. However, it is technically challenging to create such a comprehensive array encompassing all of the proteins encoded by about 23,000 genes in the human genome with current cloning and expression technologies. The most comprehensive protein microarray currently available commercially, trademarked as ProtoArray® by Life Technologies (Carlsbad, CA, http://www.lifetechnologies.com), consists of only 9,000 human proteins that have been expressed and purified from a baculovirus-based expression system. Moreover, issues with protein conformation, autophosphorylation, and stability on the array surface, as well as complications in data interpretation as a result of the presence of multiple phosphorylation sites within a protein potentially targeted by different kinases, have limited the practicality of this approach.

As a result, a number of peptide-based strategies have been devised including peptide libraries and peptide arrays, based on the notion that the substrate specificity of the kinase is largely defined by the flanking linear amino acid sequences around its target phosphorylation site(s) on the substrates. Based on the consensus recognition sequences for protein kinases derived from the peptide-based studies, one can deduce potential physiological substrates for each of the kinases in a proteome, coupled with information about protein–protein interactions, subcellular colocalization, and correlations in expression or activation.

Back in 1995, Luo and colleagues originally used the peptide array approach to identify and optimize substrate sequences for protein kinase A (PKA) and transforming growth factor (TGF) β receptors (Luo et al. 1995). Since then, the substrate specificities for a number of protein kinases have been elucidated and refined sequentially using peptide arrays (Schutkowski et al. 2005). There are two main strategies to be deployed for determining consensus phosphorylation sequences for protein kinases. On the one hand, peptide macroarrays featuring combinatorial peptide libraries or random peptide libraries such as those on the SPOT cellulose membranes were indispensable tools for elucidating the recognition sequences targeted by the kinases for which little information on their physiological substrates is available. The availability of expanding collections of recombinant active protein kinases in the past several years has facilitated the effort in this front. On the other hand, incorporation of increasing numbers of physiological phosphorylation sites uncovered through recent large-scale mass spectrometry-based phosphoproteomics studies into peptide microarrays has also significantly improved the efficiency of the substrate peptide screening process.

Currently, many large-scale peptide microarrays comprising a large number of experimentally verified phosphosites as well as those identified and optimized using the peptide library approach are readily available through various commercial sources such as PepStar™ from JPT Peptide Technologies (JPT Peptide Technologies GmbH, Berlin, Germany, http://www.jpt.com), PamChip® from PamGene (PamGene International B.V., Hertogenbosch, The Netherlands, http://www.pamgene.com), and PepChip™ from Pepscan Presto (Pepscan Presto, Lelystad, The Netherlands, http://www.pepscanpresto.com), either as products or services, for profiling kinase activities in biological samples. Various experimental protocols based on the same approach have been developed (Schutkowski et al. 2005; Thiele et al. 2010). The approach was also adapted to kineome activity profiling in bovine samples by utilizing information gathered through bioinformatics analysis of the phosphorylation sites conserved in evolution (Jalal et al. 2009).

However, inferring endogenous kinase activities based on the data from such peptide microarrays is less straightforward than initially thought. One of the main issues associated with the current approach is the overlapping specificity among protein kinases dictated by the promiscuity in substrate recognition, especially for the kinases from the same or related families. Phosphorylation of a peptide on each spot may represent the sum of activity of all the kinases targeting this particular peptide. Indeed, a specific phosphosite sequence may be optimized through evolution to be recognized by a panel of kinases and phosphatases and not be optimized for an individual kinase or phosphatase. Thus, under most circumstances, it is impossible to directly correlate the level of peptide phosphorylation on the array with the activity of a specific kinase. It is even more challenging when the activity of kinases in crude cell or tissue lysates is to be assessed, where the cell compartmentalization has been destroyed and the proper subcellular localization of proteins cannot be maintained.

In light of these challenges, we set out to identify the optimal peptide substrate sequences unique to each kinase by combining the high-throughput capability of peptide microarrays with the power of a proprietary kinase–substrate prediction algorithm developed at Kinexus (Fig. 7.1). The algorithm was built based on the information gathered through manual analysis of close to 10,000 confirmed kinase–substrate pairs for 229 typical kinases. Coupling with the alignment of the primary amino acid sequences of the catalytic domains of protein kinases, the specificity-determining residues (SDRs) were identified, and the position-specific scoring matrix (PSSM) was generated for each of the kinases for predicting their respective recognition sequences around the phosphorylation site. The PSSMs were then used to derive the optimal substrate peptide sequences. In total, 445 15-mer peptides corresponding to the predicted sequences with a single phosphorylatable residue (Ser, Thr, or Tyr) in the middle were synthesized and immobilized onto an epoxysilane-coated glass microarray surface. The resulting peptide microarray was made up of four identical subfields to allowing four kinase assays to be run in parallel. Phosphorylation of the peptides on the array was carried out by applying active protein kinases individually into each field under their respective assay conditions, and the extent of peptide phosphorylation was then detected with Pro-Q Diamond (Life Technologies), a fluorescent dye that had been validated to bind specifically to phosphorylated residues including Ser, Thr, and Tyr, regardless the context of sequences they are in. So far, over 200 protein kinases have been assayed. Many highly reactive and selective peptides have been identified as substrates for the kinases tested. While most of the sequences conformed to those reported previously, some novel motifs were also uncovered. Detailed analysis of the "hit" peptide sequences is expected to reveal the prototype optimal substrate peptides unique to each kinase, which will then be further optimized for their reactivity and selectivity using an oriented peptide library approach. It is expected in the near future a peptide microarray spotted with substrate peptides that are preoptimized to each of the kinases will become available from Kinexus for kineome profiling in complex biological samples.

Protein Phosphatase Activity Fingerprinting

Compared to protein kinases, protein phosphatases have been less well characterized with respect to their regulation and physiological substrates. This can be attributed to the misconception that phosphatases are promiscuous in substrate recognition and

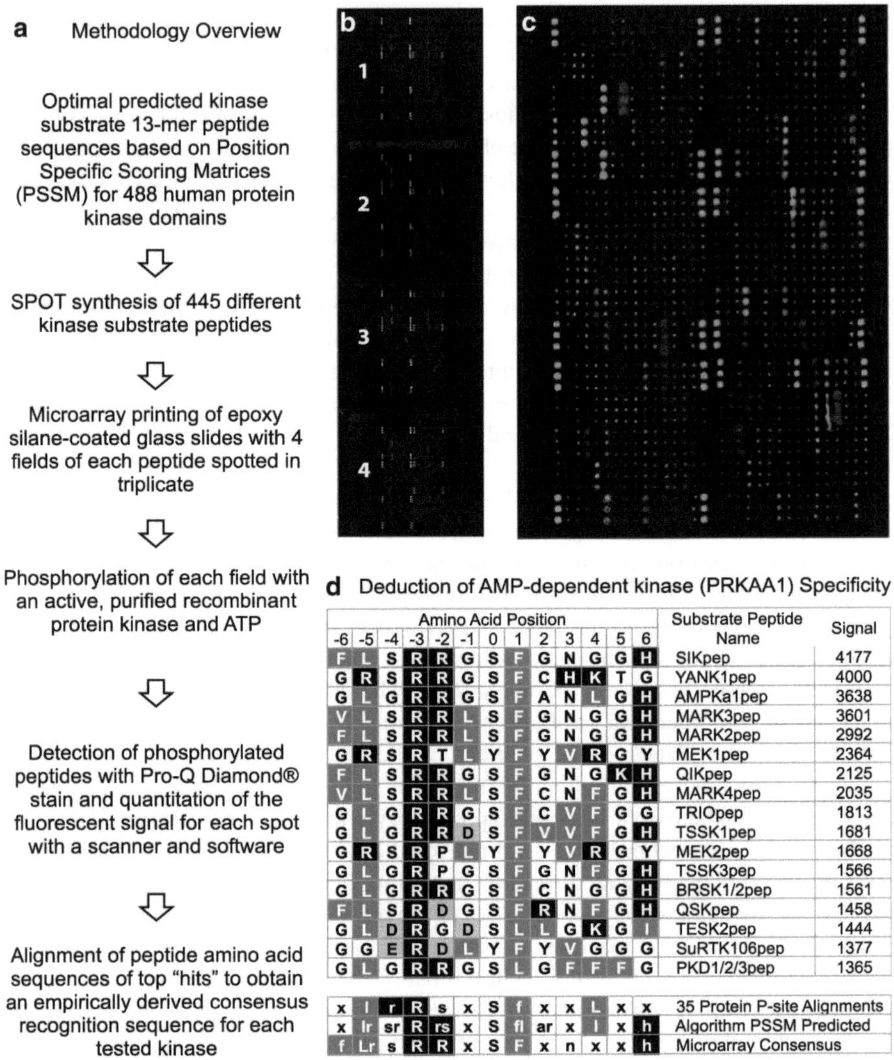

Amino Acid Position													Substrate Peptide Name	Signal
-6	-5	-4	-3	-2	-1	0	1	2	3	4	5	6		
F	L	S	R	R	G	S	F	G	N	G	G	H	SIKpep	4177
G	R	S	R	R	G	S	F	C	H	K	T	G	YANK1pep	4000
G	L	G	R	R	G	S	F	A	N	L	G	H	AMPKa1pep	3638
V	L	S	R	R	L	S	F	G	N	G	G	H	MARK3pep	3601
F	L	S	R	R	L	S	F	G	N	G	G	H	MARK2pep	2992
G	R	S	R	T	L	Y	F	Y	V	R	G	Y	MEK1pep	2364
F	L	S	R	R	G	S	F	G	N	G	K	H	QIKpep	2125
V	L	S	R	R	L	S	F	C	N	F	G	H	MARK4pep	2035
G	L	G	R	R	G	S	F	C	V	F	G	G	TRIOpep	1813
G	L	G	R	R	D	S	F	V	V	F	G	H	TSSK1pep	1681
G	R	S	R	P	L	Y	F	Y	V	R	G	Y	MEK2pep	1668
G	L	G	R	P	G	S	F	G	N	F	G	H	TSSK3pep	1566
G	L	G	R	R	G	S	F	C	N	G	G	H	BRSK1/2pep	1561
F	L	S	R	D	G	S	F	R	N	F	G	H	QSKpep	1458
G	L	D	R	G	D	S	L	L	G	K	G	I	TESK2pep	1444
G	G	E	R	D	L	Y	F	Y	V	G	G	G	SuRTK106pep	1377
G	L	G	R	R	G	S	L	G	F	F	F	G	PKD1/2/3pep	1365
x	l	r	R	s	x	S	f	x	x	L	x	x	35 Protein P-site Alignments	
x	lr	sr	R	rs	x	S	fl	ar	x	l	x	h	Algorithm PSSM Predicted	
f	Lr	s	R	R	x	S	F	x	n	x	x	h	Microarray Consensus	

Fig. 7.1 A bioinformatics algorithm-guided identification of the optimal peptide substrates unique to each kinase on peptide microarrays. *Panel* a. Schematic description of the workflow from peptide substrate sequence prediction using the Kinase Predictor 1.0 Algorithm developed by Kinexus to select test peptides for phosphorylation by kinases on the peptide microarray and, finally, to the deduction of the optimal substrate sequences for individual kinases. *Panel b.* Scanned image of the full Kinex™ Kinase Substrate Peptide Microarray phosphorylated with three different kinases. The second field was incubated with ATP in the absence of added kinase as a control. Each peptide featured a phosphorylatable residue (Ser, Thr, or Tyr) in the middle. The strong spots common among all four fields are the orientation markers designed for easy peptide localization. *Panel* c. Close-up scanned image of one field of the Kinex™ Kinase Substrate Peptide Microarray. *Panel* d. Alignment of the top phosphorylated peptides detected following incubation with AMP-dependent protein kinase alpha 1. Peptides were ranked according to their respective phosphorylation signal intensity, and an optimal substrate peptide sequence is shown in the *bottom row*

regulated in less stringent fashion in vivo, which might have arisen from that observation that a relatively small number of protein-serine/threonine- (Ser/Thr-) specific phosphatases are able to catalyze a myriad of dephosphorylation events, and that most protein phosphatases have not been found to recognize well-defined linear sequences or consensus motifs within their substrates so far.

Despite prevailing evidence that short synthetic phosphopeptides are poor phosphatase substrates compared to their parent proteins (Zhao and Lee 1997), as supported by the role of regulatory subunits in forming the substrate-binding sites required for substrate recognition according to crystallography studies (Virshup and Shenolikar 2009), several phosphopeptide-based studies have been reported that aimed at the delineation of substrate preferences using either activity- or interaction-based approaches (Sun et al. 2009; Wang et al. 2002).

Among the two main classes of protein phosphatases, protein-tyrosine (Tyr-) phosphatases (PTPs), not protein-Ser/Thr phosphatases, had been the focus of early studies on substrate specificities, due to the availability of better characterized phospho-Tyr antibodies than phospho-Ser/Thr antibodies. In those studies, phosphatase substrate specificities were commonly delineated using individually synthesized phosphopeptides (Cho et al. 1993; Zhang et al. 1994). In recent years, peptide arrays, peptide microarrays in particular, have been demonstrated for their utility in protein phosphatase specificity mapping and activity profiling.

In 2008, Waldmann's and Yao's groups independently used phosphopeptide microarrays for large-scale, high-throughput characterization of PTP and protein-Ser/Thr phosphatase substrate specificities, respectively (Köhn et al. 2007; Sun, et al. 2008), for the first time. While a fluorescently labeled phospho-Tyr antibody was employed to monitor dephosphorylation of tyrosine in the peptides in Waldman's study, Yao and coworkers used Pro-Q Diamond dye to detect dephosphorylation of Ser/Thr, circumventing the detection problem as a result of the lack of well-characterized generic antibodies for phospho-Ser/Thr. The dye has recently extended to the detection of dephosphorylation of tyrosine in place of anti-phospho-Tyr antibodies on peptide microarray by the same group (Gao et al. 2010). A phosphopeptide microarray featuring the most evolutionarily conserved human phosphorylation sites is now being explored for its potential for the determination of phosphatase specificities and activity profiling in our laboratory.

Protease Activity Profiling

In addition to protein kinases and phosphatases, peptide microarrays have also been successfully used to characterize protease specificity, based on the notion that proteolytic cleavage can be monitored by the changes in fluorescent signals on fluorogenic peptides immobilized on the array upon the action of proteases. Salisbury et al. (2002) used a fluorogenic peptide substrate array with 361 spatially addressable peptides to decipher the specificity of thrombin. Gosalia and colleagues employed a solution-phase fluorogenic peptide microarray, in which peptides were spotted as spatially separate nanodroplets, to reveal the evolutionary conservation of substrate specificity of thrombin from human, bovine, and salmon (Gosalia and Diamond 2003). The approach was also applied to determine substrate preferences of 13 serine and 11 cysteine proteases (Gosalia et al. 2005). Winssinger et al. (2004) generated a library of 192 peptides tagged with peptide nucleic acid (PNA) molecules and incubated it with protease in solution, followed by spatial deconvolution on a DNA microarray to profile the substrate specificities of thrombin, plasmin, and caspase-3.

The peptide array-based protease specificity profiling approach has now become an essential part of protease characterization platform, complementary and synergistic to other proteomic approaches used to detect alterations of substrate abundance and to identify and quantitate proteolytically generated neo amino- or carboxy-termini (auf dem Keller and Schilling 2010).

7.1.3.2 Domain-Mediated Protein–Protein Interaction Mapping

The application of peptide arrays for protein–protein interaction characterization has been well documented since the advent of SPOT peptide synthesis. It is applicable to characterizing protein–protein interactions where the interface between the two interacting proteins can be recapitulated by linear peptide sequences derived from the parent proteins. It is even more advantageous compared to other proteomics techniques such as protein arrays when the protein–protein interactions mediated by PTMs such as phosphorylation are concerned, as amino acids carrying corresponding PTMs can be readily incorporated into specific sites during peptide synthesis. Not only can the peptide array-based approach be used to map the consensus sequences recognized by these domains, it can also provide dynamic information on signal-dependent change in molecular networks for proteins defined by the peptides on the array and the proteins for which the binding is monitored (Sinzinger and Brock 2010).

Intracellular signal networks are organized through the interactions of proteins, which are often mediated by a group of diverse modular protein interaction domains (PIDs) with defined specificity. Among them, the Src homology 2 (SH2) domains are the largest family recognizing tyrosine phosphorylated sequences, and thus play pivotal roles in relaying information flow emanating from receptor protein-Tyr kinases (Pawson 2007). A phospho-Tyr-oriented peptide library with only one amino acid introduced at the defined positions at a time, and a mixture of amino acids at the randomized position, was spotted on the array and was interrogated with 120 bacterially expressed human SH2 domains, and the phosphotyrosine-containing peptide sequence motifs for 76 of them were defined (Huang et al. 2008). Combining the power of phage-displayed libraries, SPOT technology, and bioinformatics, the peptide array-based approach was also successfully used to deduce the consensus sequences of yeast SH3 domains (Tonikian et al. 2008, 2009). A peptide microarray featuring peptides with inverted configuration representing 6,223 C-terminal sequences of human proteins was probed with a PDZ domain to screen for putative interaction partners (Boisguerin et al. 2004).

With the knowledge of consensus sequences for the PIDs, peptide microarrays that carry peptides corresponding to known sequences recognized by SH2, SH3, PDZ, and other PIDs have been employed to profile the binding of proteins from complex biological samples to detect the differences in molecular interactions between different physiological states (Stoevesandt et al. 2005; Sinzinger and Brock 2010).

A peptide microarray populated with peptides, in which kinase consensus sequences and caspase cleavage recognition motifs (identified through a search of the human proteome) are overlapping, was employed in a study to investigate the role of phosphorylation in the regulation of caspase signaling pathways. Protein kinase CK2 emerged as the kinase with the most number of substrates that contained kinase consensus sequences that overlapped with caspase-3 cleavage motifs, indicating a role of phosphorylation in the inhibition of caspase-mediated apoptosis signaling pathways (Duncan et al. 2011).

7.1.4 Potential of Peptide Arrays in Clinical Applications

As a natural extension to their classical application in antigenic epitope mapping, peptide arrays displaying a collection of biologically active synthetic peptides have been demonstrated in recent years to be a very versatile tool for profiling the antibody repertoire in complex biological samples such as serum, urine, saliva, and other types of body fluids for the diagnosis of pathogen infections, allergy reactions, and autoimmunity, based on the notion that the immune response to pathogens, allergens, or autoantigens can be captured by the presence or absence of specific populations of antibodies. Hence, serological mapping has become one of the most sought after applications of the peptide array technology, as it appears to have the greatest clinical potential. Peptide libraries featuring fragments derived from autoantigens, allergens, or viral proteins presented on either the SPOT membrane-based peptide macroarrays or glass slide-based peptide microarrays have been used for antibody profiling in clinical samples. The clinical potential of such analyses have been shown by their use for antibody spectrum profiling in the sera from patients infected with hepatitis B and C, with a simian–human immunodeficiency virus (SHIV), the severe acute respiratory syndrome (SARS) corona virus, and with herpes virus. This provides crucial information not only for infection diagnostics but also for the development of vaccines (Neuman de Vegvar et al. 2003; Duburcq et al. 2004; Guo et al. 2004; Andresen and Grötzinger 2009).

The use of peptide arrays in kineome profiling has also inspired the exploration of their application in the studies of human diseases. As increasing numbers of kinase substrate peptides have been identified in recent years, peptide microarrays with the capability of screening a broad range of protein kinases have been established and used to profile the aberrant kinase activity in clinical samples as well as for monitoring the response to kinase inhibitory compounds in a high-throughput manner. This underscores the potential of peptide arrays in disease diagnosis and drug discovery (Piersma et al. 2010). Among the handful of studies reported so far in this area, Schrage et al. (2009) recently reported the activation of multiple pathways in relation to AKT/GSK3β, PDGFRB, and Src protein kinases in chondrosarcoma cells on a kinase substrate peptide array containing 1,024 peptides. Supplemented with the cell viability data in vitro, the study indicated that the Src inhibitor dasatinib is a potential treatment option for patients who are inoperable (Schrage et al. 2009). Tuynman and colleagues investigated the molecular mechanism underlying anticarcinogenic activity of celecoxib (Celebrex), a selective cyclooxygenase-2 (COX-2) inhibitor, against colorectal cancer (CRC) using a kinase substrate peptide array with 1,176 different kinase substrate consensus sequences and found that celecoxib represses c-Met-dependent signaling, which in turn led to downregulation of oncogenic Wnt signaling in CRC, supporting the potential of targeting c-Met and Wnt signaling in CRC therapy (Tuynman et al. 2008).

Recently, a cellulose membrane-based peptide array of 70 peptides derived from p160 peptide, a cancer cell targeting peptide identified by phage display, was employed to optimize the affinity of the peptides for human cancer cells using peptide-whole cell interaction assay (Ahmed et al. 2010). The binding of the three peptides with the highest affinity and selectivity for cancer cells was further confirmed using fluorescence imaging and flow cytometry. The study revealed the potential of the peptide array-based whole cell binding assay for screening and identifying cancer cell targeting peptides for cancer diagnosis and drug targeted delivery.

7.2 Peptide Arrays for Kinase Activity Assays

Rob Ruijtenbeek

7.2.1 Application Fields for Peptide Arrays Measuring Activity

7.2.1.1 Functional Proteomics

Peptide microarrays broaden a new field of research and applications often referred to as functional proteomics (Thiele et al. 2011). While DNA and protein arrays mostly focus on determination of abundance of RNA or protein molecules, peptide arrays allow the *functional* analysis of multiple proteins or protein families (Fig. 7.2). By functional analysis, we mean the detection of protein activity. Clear examples are the detection of enzymatic activities, for example, of kinases, phosphatases, and proteases in lysates from cells or tissues. However, nonenzymatic functions, like the responses to hormone binding of nuclear receptors in terms of specific coregulator protein recruitment, are also currently being studied on peptide arrays (Heneweer et al. 2007; Koppen et al. 2009). Peptide arrays enable the miniaturization and multiplexing of activity-based assays.

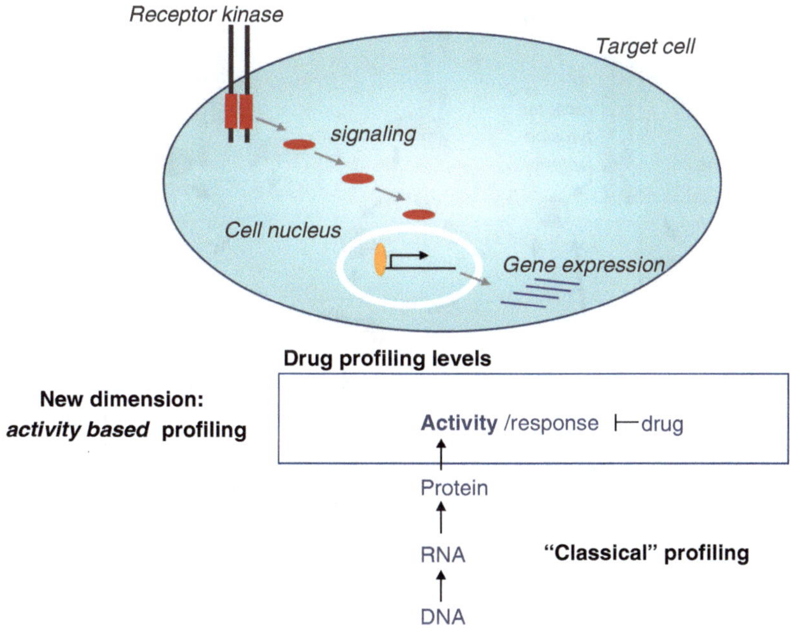

Fig. 7.2 Molecular profiling aimed at identification of molecular biomarkers often involves detection of DNA mutations, or measurements of RNA or protein abundance levels. Functional proteomics approaches, however, profile the *activities* of proteins instead. Kinase activity profiling is an example in which the enzymatic activities of kinases, playing a central role in signal transduction—in many cases leading to gene and protein expression—are measured

In the context of pharmaceutical research, and in the field of translational medicine in particular, such array-based approaches are emerging. This makes sense since the majority of the drugs being developed target protein activity and function. These new and more targeted drugs act by effecting protein function rather than targeting DNA or RNA or interfering with the modulation of protein levels. Because functional profiling of the interaction of drugs with cellular or tissue samples is of specific interest in pharmaceutical research, peptide microarrays are proving to be very useful with their ability to profile protein activity and its modulation by drugs.

7.2.1.2 Kinase Drug Targets

We focus here on the drug class of kinase inhibitors which have been reshaping the oncology field due to their high success rate. These molecules inhibit kinase function by reducing kinase activity, which can be monitored on a peptide array.

Kinases play a pivotal role in cellular biology by being the key regulators of signal transduction. Signals being detected by a membrane-bound receptor are transduced to the inner parts of the cell to result in an appropriate response.

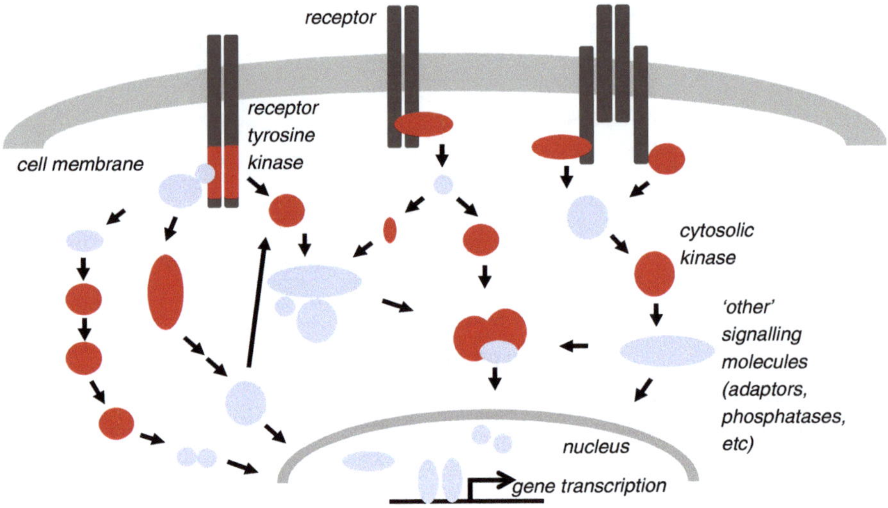

Fig. 7.3 Kinases in signal transduction cascades. In this figure, the complexity of signal transduction is represented by the interaction of multiple proteins of which the kinases are shown in *red*

This happens via highly complex cascades of events in which the signal is received and propagated using transphosphorylation reactions (Fig. 7.3). These reactions are catalyzed by protein kinases together with the crucial ATP molecule. ATP is important as not only does it provide the kinase's energy source, it also supplies the phosphate moiety, vital to the whole signal transduction cascade. A kinase becomes activated and places this phosphate group on a substrate protein; this being the subsequent link in the signal transduction pathway. Often this substrate protein is a kinase as well. A signal transduction event can be compared to a relay in athletics, where each kinase gets activated by an upstream event and subsequently passes on the baton to the next member downstream in the pathway. Most protein kinases have distinct preferences for the aromatic hydroxyl groups of tyrosine residues or for the aliphatic serines or threonines. It is this characteristic which divides this family of more than 500 members into two kinase subfamilies: protein-tyrosine kinases (PTKs) and protein-serine/threonine kinases (STKs).

7.2.2 *Technology and Applications*

7.2.2.1 Kinase Activity Profiling Technology

While in classic kinase assays the activity is detected by the phosphorylation of a single substrate, multiple substrates can now be immobilized and monitored on a microarray. Instead of placing multiple protein substrates on a chip, only the phosphosites (the sites within the protein which become phosphorylated) are immobilized in a peptide microarray. Thus, the peptides represent the protein substrates. As has been discussed in the previous chapters, this can be done in a variety of ways, but all are based on a solid support. In most cases, the sequences are derived from known phosphorylation sites in the human proteome. As the human proteome is estimated to comprise more than a million proteins, of which more than two-thirds can be phosphorylated, this indicates the huge amount of different phosphosites that can be investigated by peptide arrays.

The principle of the assay is that the kinase activities in the sample of interest phosphorylate the peptides. The phosphorylation event is detected by either radiography or fluorescence imaging of the array. In radio assays, the peptide is phosphorylated using radioactive ATP as the phosphate source. This approach is increasingly being replaced by the use of fluorescence assays. In the latter case, the phosphorylation of the peptide is detected by a fluorescently labeled molecule which is either a chelator (e.g., phosphotag) or an antibody. Ideally, the antibody needs to detect the phosphoamino acid in all the available peptide sequences on the chip equally well and independently of the adjacent amino acids. Antibodies like PY20 work very well in detecting tyrosine phosphorylated peptides, but for serine/threonine phosphorylated peptides, cocktails are needed for full coverage of detection.

The first peptide microarray applications for kinase profiling used glass as a solid support and radioactivity for readout at a single time point. Later, protocols were developed based on the fluorescent readout of labeled antibodies (or cocktails of antibodies) binding to phosphorylated peptides. A second generation of this technology was developed by researchers in the Netherlands and is referred to as the PamChip® technology (Lemeer et al. 2007; Hilhorst et al. 2009; Versele et al. 2009) (Fig. 7.4). With this technology, antibody-based fluorescence detection has been combined with a change of solid support from glass slides to a porous ceramic. In this format, the sample is pumped up and down through the porous aluminum oxide ceramic sheets, in which the peptides are immobilized at designated spots. Each spot comprises thousands of separated pores with diameters of 0.2 μm in which the peptides are site-specifically immobilized. Each time the sample is below the solid support, the degree of phosphorylation is monitored by imaging the fluorescence intensities caused by the antibody binding to the phosphorylated peptides alone. These time curves, or kinetic readouts, appear to be instrumental in the enzymatic studies; a kinase is after all an enzyme which catalyzes the rate (the kinetics) of a phosphorylation reaction. In addition, the kinetic and multi(time) step readout for each of the 144 or 256 peptides on each single array allows much more comprehensive statistical analysis of the signals than data from a single time point per peptide spot on a glass array (Thilakarathne et al. 2011).

The application of peptide arrays in biological, pharmaceutical, and medical studies often requires the analysis of many samples under variable conditions. For example, lysates from cells should be analyzed using a range of time points, varying concentrations, and with multiple different drugs. For this reason, a system has been developed which has the capability of analyzing 96 arrays at once. This latest technology for kinase activity profiling is based on a 96-well plate format, in

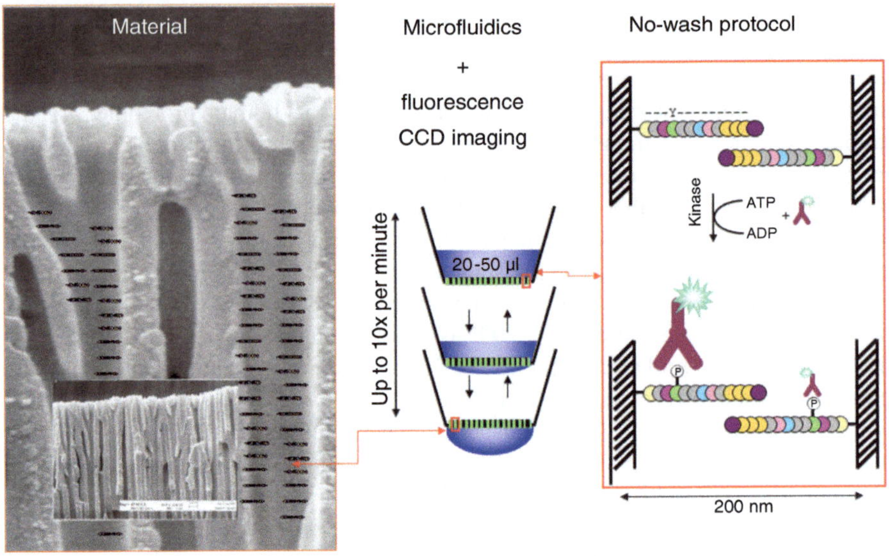

Fig. 7.4 PamChip® technology is based on a porous substrate made of aluminum oxide, in which pores the peptides are immobilized (*left panel*). Due to this porosity, the sample can be pumped up and down through this solid support. Every time when the sample is positioned below the microarray, an image is taken of the microarray by a CCD camera (*middle panel*). Via this real-time imaging of the microarray, the signal, developing in the peptide spots due to binding of fluorescent antibody detecting peptides phosphorylated by the kinases, can be monitored

which each well comprises a peptide microarray. Bioinformatics for analysis of the vast datasets from such studies has been evolving in parallel. Thilakarathne et al. (2011) developed a new method based on semiparametric mixed linear models to further enhance the amount of information that can be obtained from the multiparallel kinetic readouts from each microarray.

7.2.2.2 Applications

A straightforward application is substrate identification using recombinant kinases. Such studies have indicated that different kinases have their own preferences for the peptide sequences they phosphorylate. Clear differentiation between the PTKs and STKs has been confirmed, although dual specificity kinases have also been found. In addition, it has become apparent that although each kinase has a preference for particular peptide sequences, they can also be promiscuous, resulting in multiple peptides being phosphorylated to different degrees in diverse peptide sets. In short, the degree of phosphorylation by purified kinases varies from peptide to peptide and can be profiled in hundreds per array, resulting in phosphorylation *fingerprints*.

Substrate profiling studies have revealed important biological information as described in the paper by Schutkowski et al. where they showed that for optimal recognition by GSK3b, a peptide substrate should be prephosphorylated or primed

(Schutkowski et al. 2004). Another interesting application was explored by Poot et al. who identified an optimal substrate for PKC isozymes and coupled this peptide to an ATP-binding site inhibitor to generate a bisubstrate (Poot et al. 2009; van Ameijde et al. 2010). The peptide microarrays were subsequently used to evaluate the resulting inhibitors, which were potent and selective toward the theta isozyme.

With the development of protocols for profiling cell lysates of tissue homogenates, the application area has been broadened to signal transduction and pathway studies [well reviewed by the group of Schutkowski (Thiele et al. 2011)]. The effect of a stimulus or a kinase-inhibiting drug on cultured cells can now be investigated at the complexity level of a cell, where multiple kinases can be active in the context of the interacting networks that exist. At this point, peptide arrays provide a welcome extension to classical methods like (phospho) Western blots and ELISAs, which monitor drug effects on a (single) kinase by detecting the variation in abundance of the downstream phosphorylated substrate. The peptide arrays monitor the enzyme activity of multiple kinases at once and not only the end result of this activity.

An interesting feature of functional proteomics is found in the ability to study *direct effects* of the investigative drug. Because activities of kinases can be monitored in cell lysates or tissues, drugs can be characterized in a complex, and probably more realistic, context than in classical single readout (singleplex) assays. This latter type of assay is limited as it can only investigate the activity of the isolated drug target. Drug selectivity profiling is a clear example of an application which benefits from the combination of multiplexing and miniaturization (Fig. 7.5).

Another example of an application area is the unraveling of a drug's cellular mechanism of action. In this application, the peptides on the chip represent multiple different proteins involved in complex cellular pathways and signal transduction networks. In a small lysate sample, derived from just 10,000 to 100,000 cells or less than a tenth of a cubic millimeter of tissue, multiple diverse interactions can be studied at once.

7.2.2.3 Reversed Translational Medicine and Biomarkers

A recent development of peptide microarrays has been the application of new drugs in pharmaceutical research and clinical development. In the field of oncology in particular, fundamental progress has been made by so-called *targeted medicine*. Previously, anticancer drugs were targeting cellular processes, like cell division, more globally. New insights into cell signaling and signal transduction cascades have changed the way novel oncological drugs are being developed, and it is the kinase enzyme class which is playing a crucial role in this progress. Many of its members play a pivotal role in the mechanisms of tumor genesis, and some kinases are even the active protein products of oncogenes. Examples of successful cancer drugs targeting protein kinases—or the signaling pathways they are involved with—are imatinib, erlotinib, gefitinib, and the previously mentioned sunitinib

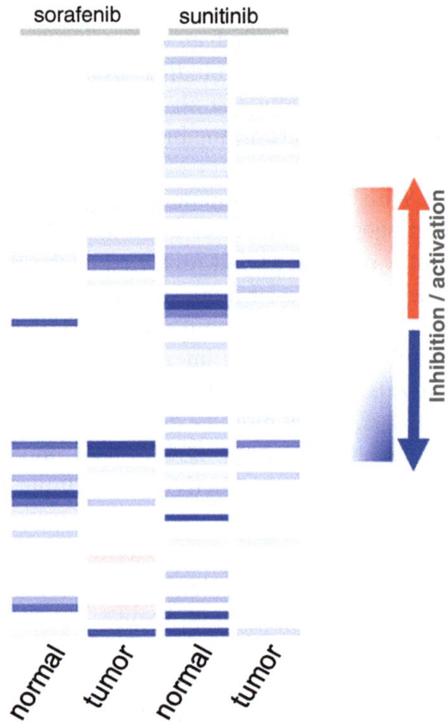

Fig. 7.5 Selectivity profiling of kinase inhibitor drugs using peptide microarrays. Here sorafenib inhibition profiles are compared with sunitinib in extracts from both normal and tumor tissue from a renal cell carcinoma patient

and sorafenib. These are all molecules that block the catalytic activity of protein kinases. A related class of therapeutics is antibodies, which intervene in a different way with cellular signaling: they act by blocking the initiation of receptor signaling. Examples of the latter are trastuzumab and bevacizumab, which block EGFR kinase and VEGFR signaling, respectively.

These drugs can be studied comprehensively with peptide arrays. In these studies, two formats are currently being used. Using cell line models, the cells are either treated with the drug in culture or on the chip. In the first case, lysates are prepared from the cells before and after treatment and profiled for activities on the chip. In the second case, lysates can be treated directly by spiking the drug into the solution just before application onto the chip. In the latter instance, cell lines, tissue homogenates from animal models, or even clinical samples can all be used. The effects of the inhibitors on the kinase activities in these samples can all be directly assessed. Although the highly important context of the cellular architecture is lost, which is surely a downside, the potential to profile all detectable, full-length kinases—with their relevant posttranslational modifications—in the same sample, opens up vast new fields of applications.

In drug discovery, researchers screen for kinase-inhibiting compounds in chemical libraries. During such studies, they often use an abstracted model, the purified protein, but this protein is frequently truncated to its domain only. A major disadvantage of this approach is the absence of other domains, including those with a regulatory function. In the recently developed protocols for peptide array analysis of kinases in cell and tissue lysates, the drug target can now be studied more naturally as a full-length protein, in the way it is actually expressed in cells or tissues. At first, this was shown in a model system, but interestingly, this approach appears to be translatable to patient-derived tissues. This means that the kinase drug targets can now be studied in the same form as they are expressed in a patient's tumor, thus full length, fully decorated with all relevant posttranslational modifications and in the presence of stabilizing or activating cofactors (e.g., heat-shock proteins). In addition, they can be studied in the presence of all other kinases expressed in the cell or tissue being investigated. Such analysis of patient-derived tumor samples can result in the identification of tumor-specific kinase activities. When linked to pathological, diagnostic, and/or clinical data, this can lead to the identification of diagnostic or prognostic biomarkers.

While the on-target effects are being monitored, the researchers can also obtain insights into the drug's effect(s) on other kinase targets, which are either intended—in the case of multitarget inhibitors—or unintended and can putatively cause side effects. The latter opens up new opportunities for the toxicologist in investigating and understanding adverse drug reactions leading to toxicological biomarkers.

A very typical feature of activity-based assays is the capability of drug testing. With peptide microarray-based kinase assays, not only can multiple kinases in a patient sample be studied at once, their response to their inhibiting drugs can also be studied. This possibility links the presence and activity of the kinase drug targets to their responsiveness to the drug.

7.2.2.4 Personalized or Precision Medicine and Drug Response Prediction

Drug response is a leading parameter in the clinical development of a drug. In the development of kinase inhibitors in cancer, the response rates are often very low, even in case of effective drugs. These drugs are developed against specific kinase targets, but these targets are not always equally present or active in the whole treated population. Furthermore, in a subset of nonsensitive or resistant patients, the role of this target in tumorigenesis and growth or metastasis is not essential and can be overruled by other mechanisms. In order to identify the patient subpopulation that is likely to respond, tests need to be developed that match the right patient to the right drug and vice versa as more drugs are being developed.

There are already examples of such companion diagnostic tests. For the prediction of response to trastuzumab (Moelans et al. 2011), targeting the receptor tyrosine kinase Her2/Neu, patients are tested for the presence of this target on their tumor cells, before they receive this breast cancer therapy. A recent example is the test for the ELM4-ALK translocation to select patients for pharmacotherapy with crizotinib (Kwak et al. 2010), an ALK kinase inhibitor. If the whole lung cancer patient population would have been treated, only an extremely low percentage would have shown a clinical response because only 4% have this mutation. The availability of the companion diagnostic test was therefore essential for the success of the clinical trial.

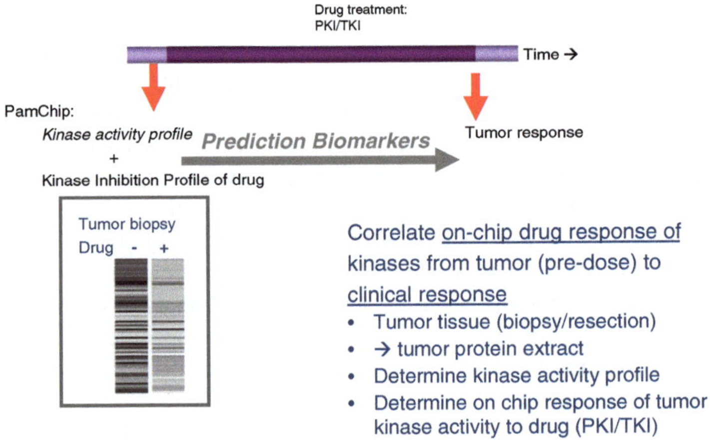

Fig. 7.6 Strategy for prediction of a patient's drug response using pretreatment biopsies on kinase substrate microarrays

The identification of predictive biomarkers appears to become essential in many drug development programs. The classical technologies for biomarker discovery are based on testing for DNA mutation or RNA or protein expression levels. Molecular data are obtained in biopsies taken before the patient is treated. If these data can be correlated or associated to the therapy response, this can be the start of generating a companion diagnostic test. Peptide microarrays are also currently being used in this effort. While classical methods cannot involve the drug of interest in predose tissue samples, kinase microarrays can, as discussed above. In addition, the drug can actually be used in the test. This means that drug-specific data and information can be generated using predose biopsies. Proof of concept of this approach was shown by Versele et al. in a multiple cell line study. Analogous to the way it is aimed to work in a clinical setting, they profiled the lysate of a cell line on a peptide microarray in the presence and absence of their drug candidate. The inhibition profiles were used to predict the response of the cell line to drug treatment in culture. From these profiles, they could identify a set of peptide phosphorylations of which the response (inhibition) on the chip was predictive for the tumor cell proliferation (Versele et al. 2009). This concept (Fig. 7.6) is now being explored in clinical studies by my research group in collaboration with the Netherlands Cancer Institute (NKI), the VU Medical Center, and other cancer centers in both the USA and Japan. In a study presented at ASCO in 2011 on neoadjuvant treatment of non-small-cell lung cancer with the EGFR kinase inhibitor erlotinib, we showed that candidate biomarkers could be identified. On-chip peptide phosphorylations and inhibitions were correlated to clinical responses. With no information on the pathological assessment of the resection tissues available to the testers, a model built on those profiles could still predict the pathological response (Hilhorst et al. 2011). It should be noted that resection tissue was used and not pretreatment biopsies which is needed to make this into a companion diagnostic test. Nonetheless, this shows a promising new application of peptide microarrays.

7.2.3 Conclusion and Future Applications in Personalized Medicine

It could be possible to apply this principle of drug testing on patient-derived tissues to other targeted therapies as well. In addition, if other protein classes are targeted, for example, phosphatases, proteases, nuclear receptors, acetyltransferases, histone deacetylases, and methyltransferases, the target responses in patient-derived samples could be tested using a peptide microarray.

The nonfocused, nonbiased, and global profiling nature of the arrays allows parallel monitoring of drug targets and class-related nontargets. These nontargets can be functional proteins involved in the mechanisms of resistance and are therefore possibly very useful markers for predicting resistance to targeted therapies.

Finally, nontargeted therapies such as chemoradiation could also be accompanied in the future by such testing methods, as was shown in a recent publication by a Norwegian group. They generated kinase activity profiles of tens of biopsies taken before patients were treated and could identify peptide phosphorylation patterns that correlated to the tumor regression grade after therapy. They generated a response prediction model that could predict the responses of a newly tested set of patients with promising accuracy (Folkvord et al. 2010).

7.3 Peptide Microarrays by Laser Printing

Thomas Felgenhauer, Ralf Bischoff, Frank Breitling, and Volker Stadler

7.3.1 Introduction

Several sophisticated methods are in use worldwide to produce peptide microarrays. Each of these methods has its special advantages and drawbacks. High amounts of identical oligomers are achievable on cellulose supports via SPOT synthesis (Frank 1992; Dikmans et al. 2006), but spot densities are very low due to droplet handling. With photochemical methods where chain growth is induced by a laser beam, very small spot sizes and high spot densities are possible (Fodor et al. 1991a, b; Lipshutz et al. 1999). In this case, the drawback is in the sequential use of monomer solutions which might be acceptable in DNA synthesis (four monomers), but the yield is dramatically reduced when the number of needed coupling cycles increases as in the case of standard peptide synthesis where minimum 20 individual cycles are needed to complete a fully combinatorial layer.

The use of a laser printer as synthesis machine makes it possible to overcome the obstacles of the methods described above. Solid particles (toners) carrying the reactive building blocks are printed in parallel in high resolution to a desired support. A full combinatorial layer is developed—like a color picture printout—at once, and the coupling cycles are reduced from 20 to a single one per layer (Stadler et al. 2008).

7.3.2 Technical Aspects of Commercial Laser Printers

A commercial laser printer uses small solid toner particles (~10 μm) that are triboelectrically charged by friction inside a toner cartridge drum system. Because of the materials involved, this procedure leads to strong electrical charges on the particle surfaces, which enables the directional movement of the particles within electrical fields. A laser beam or an LED row translates 2D light patterns into electrical patterns on top of an organic photoconductor drum. These images are developed with the charged toner particles that are finally transferred to the support. At office applications, a color laser printer system delivers four different color toners (black, cyan, yellow, and magenta) on a sheet of paper with a resolution of 1,200 up to 2,400 dpi. The polymer-based toner particles are fixed to the cellulose support by heat.

7.3.3 Combinatorial Synthesis with Laser Printers

The main challenge in combinatorial synthesis is to deliver different kinds of monomers with high accuracy to their designated reaction partner or reaction site. Whereas a color laser printer delivers only four toners, a peptide synthesizer based on the xerographic technique should be able to handle at least 20 different building blocks for basic peptide synthesis or other feasible monomers for the production of peptide mimetics (amino acids in D-form, methylated, phosphorylated derivatives, nonnatural versions).

In addition to the great flexibility of the synthesizer, an exact positioning of consecutively printed layers is the basic requirement for the parallel elongation of combinatorial assembled oligomer chains. With increasingly better printing accuracy, the spot density also increases, as well as the diversity of synthesized peptides.

7.3.4 Solid Particles as Reaction Units

To benefit from the laser printer as delivery machine for monomers in combinatorial chemistry, the toner particles (delivery packages) have to be modified for this chemical purpose. In addition to their properties as solid, electrically charged particles, they also need the attributes of a solvent once melted. This change of properties happens after the particles have been addressed to their designated reaction site, where they are transformed into a liquid sphere simply by melting. Thereby, activated monomers are mobilized, which allows them to diffuse to their reaction partner for chain elongation. These very special solid/liquid characteristics of the toner particle depend on the choice of the appropriate matrix material. On the one hand, this material should withstand the harsh mechanical treatment inside the printer (e.g., friction, charging, transport); on the other hand, the liquefaction at moderate temperatures ($<100^{\circ}C$) is fundamental in order to perform as a solvent for a chemical reaction. In addition, the matrix material should protect the reactive monomers from ambient conditions during long-term storage in cartridges, and finally, the material itself must be inert toward the components inside.

Since all the different monomer particles are addressed in parallel by the printer, all the different activated amino acid derivatives within a completed layer of amino acid particles are activated at once in a single melting step. This feature is the main advantage of our technique. Washing and deprotection steps that follow after the coupling step finish the cycle, and result, if repeated, in the combinatorial synthesis of a peptide array (Fig. 7.7).

Our method uses conventional Fmoc chemistry (Chan and White 2000) and differs from the SPOT synthesis only in the solvent we employ: it is solid at room temperature, which allows for the intermittent immobilization of chemically highly activated amino acid derivatives within particles. This activation is due to a C-terminal pentafluorophenyl ester in combination with N-terminal Fmoc protection and standard side chain protecting groups. Surprisingly, when embedded in particle matrix, these very reactive chemicals proved to be stable for months at room temperature, an exception being Fmoc-arginine-OPfp that shows a decay of 4% per month. However, if compared to the much faster decay of activated arginine esters in solution, this decay is negligible.

7.3.5 Surface Coatings

The surface-coated solid support must provide free amino groups that react with preactivated amino acid derivatives. It must stand harsh conditions during peptide synthesis (solvents, bases, strong acids during final cleavage of side chain protecting groups) and postsynthesis; it must allow for the incubation of arrays

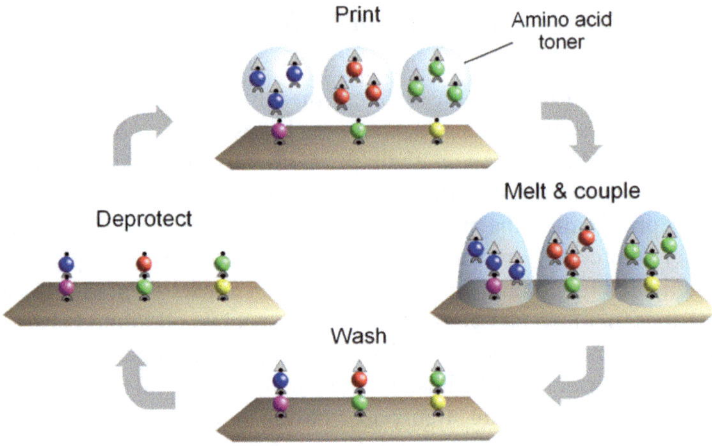

Fig. 7.7 Combinatorial synthesis with a peptide laser printer. (Print): a laser printer addresses different reactive monomers embedded in a solid matrix material in parallel to a support that displays previously synthesized peptide fragments with reactive groups. (Melt and Couple): once printed, the particles are melted. This frees the monomers to diffuse and couple to the growing chains on the support. Different reaction spheres are separated from each other due to surface tension, which constricts melted particles to small individual hemispheres. (Wash): a cycle of synthesis is completed when excessive monomers are washed away, and (Deprotect): the Fmoc protection group is removed. Repetitive coupling cycles yield a peptide array. In contrast to lithographic synthesis methods, this is done with only one coupling reaction per layer

with an analyte, for example, an antibody solution. We employ 30-100 nm thick 3D polymer coatings that have a high loading capacity (high density of amino groups). Alternatively, essentially 2D layers are used as solid supports for peptide synthesis. These are generated from functionalized silanes that also stand the conditions during peptide synthesis and, dependent on the used assay system, sometimes perform better when compared to the polymer coating. Such surfaces are described in detail elsewhere (Beyer et al. 2006; Stadler et al. 2007).

7.3.6 The Peptide Laser Printer

The peptide laser printer at PEPperPRINT GmbH (see Fig. 7.8a, b) has 24 printing units assembled in a row. Twenty of these toner cartridges that are based on the Oki system are equipped with Fmoc-amino acid esters; the four remaining cartridges are used for nonstandard amino acid particles. The printer works with micron resolution which currently allows for the synthesis of 270,000 peptides on a (20 $\times$ 20) cm^2 glass substrate. This corresponds to a spot density of ~800 spots per cm^2. Currently, the machine is used for the synthesis of up to 20meric peptides.

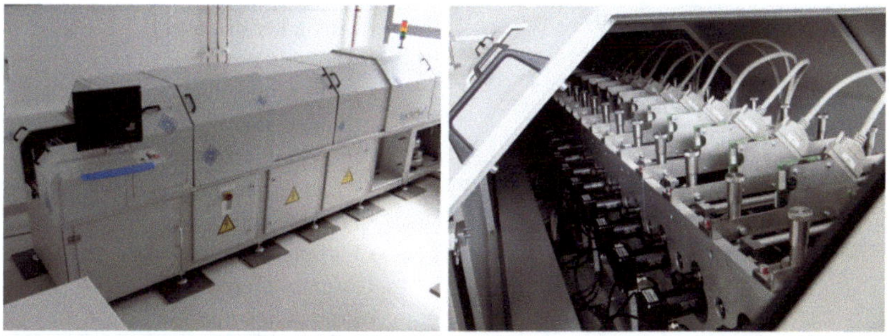

Fig. 7.8 (**a**) The peptide laser printer. (**b**) Horizontal setup of the 24 development units (printing cartridges) assembled in a row

7.3.7 Applications

Our particle-based synthesis method makes available to the scientific community, for the first time, very high-density peptide arrays. This is, in technical terms, the main novelty of this method (Table 7.1). As such, this method certainly soon will approach the number of different molecules that Nature's screening systems employ. These use, for example, millions of randomly generated antibodies to screen for binders against virtually any target molecule, among them nonnatural molecules that have never been encountered by evolution. Peptide arrays with *natural* amino acids (L-form) have been used in the past mainly as protein fragment libraries in proteome research, or as diagnostic tools for serum screening and for antibody profiling. For the development of novel therapeutics, often *nonnatural* amino acids (e.g., D-amino acids) were integrated at critical positions within the peptide sequence in order to increase the metabolic stability of the peptides. However, due to the use of only low-density peptide arrays, up to now, such screens usually had to rely on extensive previously available knowledge. Typically, a peptide sequence that was already known to bind to the target was then modified to improve the stability or the binding affinity, or an already known antigenic protein sequence was used to generate many overlapping peptides in order to narrow down an antibody's binding epitope.

In the near future, we will certainly see that very high-density and affordable peptide arrays will be used to find binders without extensive previous knowledge about the sequence of a potential binder. Similar to Nature's screening systems, a vast number of different peptides should be sufficient to find binders against nearly any target molecule, and different from surface display techniques, the array format will allow for an easy and unequivocal discrimination of specific from nonspecific binders. Very high-density peptide arrays will be used for such screens that have been cleared from all those peptides that were found to bind to more than a few different target proteins. It is practically impossible to avoid such nonspecific binders in all the surface display methods. Thus, the high combinatorial diversity given by the laser printer method should increase the possibility to discover potent

Table 7.1 Peptides are covalently bound to a glass slide microarray substrate and available for a wide variety of binding assays including:

• Biomarker discovery
• Antibody profiling
• Epitope mapping
• Immunoassays
• Protein interactions
• Kinase screening
• Protease screening
• Phosphatase screening
• Affinity profiling
• Protein fingerprinting

lead structures. Moreover, fast follow-up synthesis of all possible permutations of these structures due to affinity maturation should reduce the timescale and the costs in *lead-to-hit* development dramatically.

7.3.7.1 Serum Profiling

Low-density peptide arrays have been used previously to narrow down the binding epitope(s) of binding antibodies by synthesizing many overlapping peptides derived from the sequence of a known antigen that has been used, for example, to immunize an animal; however, such experiments used to be prohibitively expensive. High-density peptide arrays are cheaper—they simply allow more of these experiments. It should be feasible in the future, for example, to routinely monitor the kind of antibodies that evolved in a mouse that was immunized with a protein. The experimenter could then use only those mice that evolved an antibody specific for an especially interesting epitope within the protein that was used for immunization, thus saving a lot of time and money by using only selected mice for the generation of hybridomas. Shown as an example for this statement are the results obtained when we used four different rabbit sera that were immunized with protein A and closely related protein B (Fig. 7.9). Only one of two rabbits immunized with protein A or with protein B revealed that specific antibodies have been generated, while the other two rabbits did not generate protein A- or protein B-specific antibodies. The staining pattern revealed that rabbit 1 developed several antibodies that were specific for the C-terminal region of protein A, while rabbit 3 developed antibodies that targeted the N-terminal region of protein B, and when scrutinizing the peptide sequences, both productive sera did not reveal cross-reacting antibodies against determinants from closely related proteins #A and #B.

7.3.7.2 Biomarker Discovery: Follow-Up Synthesis, Permutation Screen

A biomarker is a substance that is used as an indicator of a biological state. The biomarker serves as an indicator to measure and evaluate normal or pathogenic biological processes, or pharmacologic responses to a therapeutic intervention.

Immunization with protein A		Immunization with protein B	
serum rabbit 1	serum rabbit 2	serum rabbit 3	serum rabbit 4

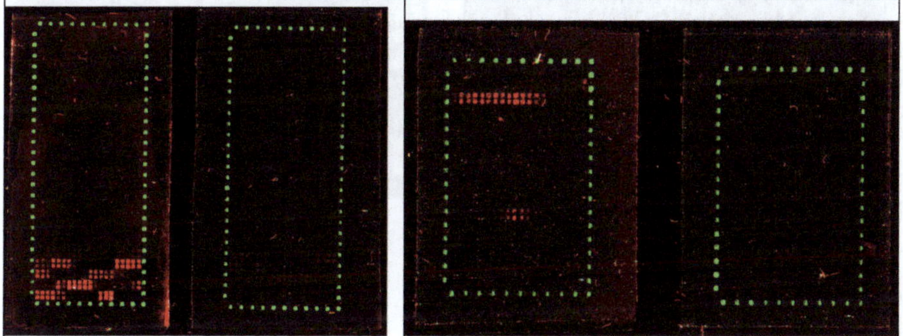

15meric peptide array of protein A		15meric peptide array of protein B	
specific response in C-terminal region rabbit 1 / protein A	Rabbit 2 developed no antibody response specific for protein A	specific response mainly in N-terminal region rabbit 3 / protein B	Rabbit 4 developed no antibody response specific for protein B

Fig. 7.9 Profiling of four rabbit sera against 15meric peptides derived from two antigens (A&B). Peptide spots are arranged in double rows and adjacent structures have an overlap of 14 amino acids. The arrays are framed by green control spots

Especially useful are biomarkers that are found in human sera. These are a rich and accessible source for the detection of diagnostic markers in human diseases. Serum antibodies have been used extensively for diagnosis, for example, of flu antibodies in a patient. Such antibodies often are found in a patient decades after infection and can easily be analyzed with peptide arrays. Especially interesting are those (peptide-)antigens (and their corresponding antibodies) that are targeted by an immune response. These are useful as biomarkers for drug discovery and diagnostics. However, the most interesting scientific question in biomarker discovery is as follows: can we find novel antigens and their corresponding antibodies without previous knowledge about the antigen? Figure 7.10a shows that such a scientific question could be answered with very high-density peptide arrays. A randomized high-density array of 15meric stochastically chosen peptides (only a detail view of ~5,000 structures is shown) was used to find peptide binders for the Flag M2 antibody with its known binding epitope NNN**DYK**NNN**D**/**E**NNN. Indeed, we could find six weak binders in a first screen. Sequences from all of these initial hits were then used in a follow-up screen that stained the completely permutated sequences from these initial binders. Figure 7.10b shows such a permutation screen that started with the sequence "ECWGDYKSMECADWH" found as an initial hit. This sequence, and all the other five hits from the initial screen, then revealed either the sequence NNN**DYK**NNN**E**NNN or NNN**DYK**NNN**D**NNN, i.e., the Flag M2 epitope. All amino acid positions depicted by "N" could be exchanged by other amino acids. It remains to be seen if such a screen could be employed to also find novel biomarkers when staining stochastically chosen very high-density peptide arrays with serum antibodies derived from patients with enigmatic diseases.

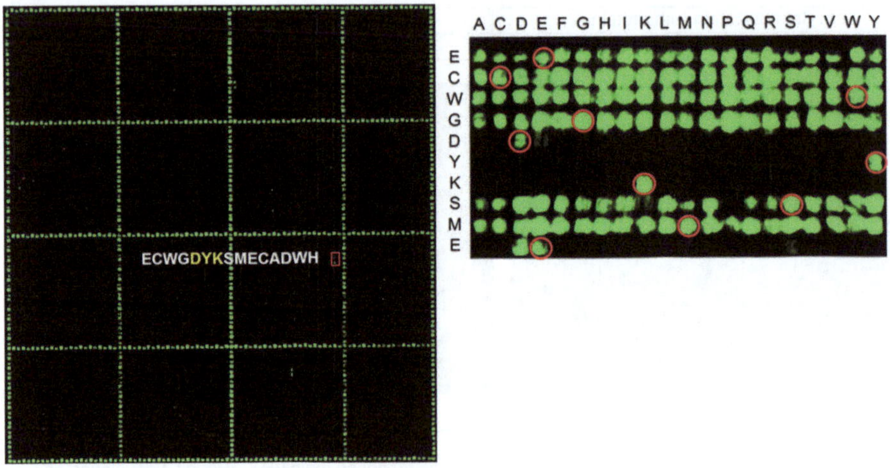

Fig. 7.10 (**a**) Lead structure "ECWG**DYK**SMECADWH" found within a randomized array of 15meric peptides. (**b**) Follow-up synthesis; full permutation of the sequence ECWG**DYK**SME. Wild-type residues are highlighted by a *red circle*

7.3.7.3 Drug Development

Many protein–protein interactions are mediated by short linear motifs. Some of these interactions are deregulated in diseases and thus are potential targets for modulating peptide-based drugs. These peptides should interfere with the protein–protein interactions by binding to one of the partners, either activating or inhibiting the signals that depend on the respective proteins. Historically, and as stated above, peptide drugs have been based, for example, upon the optimization of natural peptide hormones but, more recently, novel peptides are being developed that have been isolated from combinatorial recombinant libraries. The idea is to offer such a large number of different potential peptide-based binders that simply by chance any protein would "find" at least one binder among a plurality of different peptides. However, the current size limits of peptide arrays and the associated costs had made it, until now, unrealistic to conduct such comprehensive profiling screens.

7.3.8 Conclusion

The production of microarrays by laser printing uses a novel chemical concept where an activated monomer is encapsulated within solid particles that are sent to different "addresses" on a surface displaying reactive groups. Thereby, an established technology is used for the rapid construction of a densely spaced pattern of different kinds of particles. These comprise different building blocks that are initially "frozen" at room temperature within solid particles. Thawing these particles at once leads to a single coupling step per layer, which is the main advantage of this method when compared to sequential coupling, for example, in lithographic methods. Our particle-based method is particularly well suited for automation and, thereby, results into drastically reduced cost per peptide spot. It brings affordable high-density peptide arrays within reach of normal laboratories and may have an impact similar to the one that high-density oligonucleotide arrays had in the field of genomics.

References

Ahmed S, Mathews AS, Byeon N, Lavasanifar A, Kaur K (2010) Peptide arrays for screening cancer specific peptides. Anal Chem 82:7533–7541

Andresen H, Grötzinger C (2009) Deciphering the antibodyome—peptide arrays for serum antibody biomarker diagnostics. Curr Proteomics 6:1–12

Andresen H, Grötzinger C, Zarse K, Kreuzer OJ, Ehrentreich-Förster E, Bier FF (2006) Functional peptide microarrays for specific and sensitive antibody diagnostics. Proteomics 6:1376–1384

auf dem Keller U, Schilling O (2010) Proteomic techniques and activity-based probes for the system-wide study of proteolysis. Biochimie 92:1705–1714

Beyer M, Felgenhauer T, Bischoff FR, Breitling F, Stadler V (2006) A novel glass slide-based peptide array support with high functionality resisting non-specific protein adsorption. Biomaterials 27:3505–3514

Bhushan KR (2006) Light-directed maskless synthesis of peptide arrays using photolabile amino acid monomers. Org Biomol Chem 4:1857–1859

Boisguerin P, Leben R, Ay B, Radziwill G, Moelling K, Dong L, Volkmer-Engert R (2004) An improved method for the synthesis of cellulose membrane-bound peptides with free C termini is useful for PDZ domain binding studies. Chem Biol 11:449–459

Brognard J, Hunter T (2011) Protein kinase signaling networks in cancer. Curr Opin Genet Dev 21:4–11

Chan WC, White PD (2000) Fmoc solid phase peptide synthesis—a practical approach. Oxford University Press, Oxford, pp 41–76

Cho H, Krishnaraj R, Itoh M, Kitas E, Bannwarth W, Saito H, Walsh CT (1993) Substrate specificities of catalytic fragments of protein tyrosine phosphatases (HPTP beta, LAR, and CD45) toward phosphotyrosylpeptide substrates and thiophosphotyrosylated peptides as inhibitors. Protein Sci 2:977–984

Dikmans A, Beutling U, Schmeisser E, Thiele S, Frank R (2006) QSAR Comb Sci 25:1069–1080

Duburcq X, Olivier C, Malingue F, Desmet R, Bouzidi A, Zhou F, Auriault C, Gras-Masse H, Melnyk O (2004) Peptide-protein microarrays for the simultaneous detection of pathogen infections. Bioconjug Chem 15:307–316

Duncan JS, Turowec JP, Duncan KE, Vilk G, Wu C, Lüscher B, Li SS, Gloor GB, Litchfield DW (2011) A peptide-based target screen implicates the protein kinase CK2 in the global regulation of caspase signaling. Sci Signal 4:ra30

Fodor SP, Read JL, Pirrung MC, Stryer L, Lu AT, Solas D (1991a) Light-directed, spatially addressable parallel chemical synthesis. Science 251:767–773

Fodor SP, Read JL, Pirrung MC, Stryer L, Lu AT, Solas D (1991b) Science 251:767–773

Folkvord S, Flatmark K, Dueland S, de Wijn R, Grøholt KK, Hole KH, Nesland JM, Ruijtenbeek R, Boender PJ, Johansen M, Giercksky KE, Ree AH (2010) Prediction of response to preoperative chemoradiotherapy in rectal cancer by multiplex kinase activity profiling. Int J Radiat Oncol Biol Phys 78:555–562

Frank R (1992) Spot-synthesis: an easy technique for the positionally addressable, parallel chemical synthesis on a membrane support. Tetrahedron 48:9217–9232

Frank R, Heikens W, Heisterberg-Moutsis G, Blöcker H (1983) A new general approach for the simultaneous chemical synthesis of large numbers of oligonucleotides: segmental solid supports. Nucleic Acids Res 11:4365–4377

Gao X, Zhou X, Gulari E (2003) Light directed massively parallel on-chip synthesis of peptide arrays with t-Boc chemistry. Proteomics 3:2135–2141

Gao L, Sun H, Yao SQ (2010) Activity-based high-throughput determination of PTPs substrate specificity using a phosphopeptide microarray. Biopolymers 94:810–819

Geysen HM, Meloen RH, Barteling SJ (1984) Use of peptide synthesis to probe viral antigens for epitopes to a resolution of a single amino acid. Proc Natl Acad Sci USA 81:3998–4002

Gosalia DN, Diamond SL (2003) Printing chemical libraries on microarrays for fluid phase nanoliter reactions. Proc Natl Acad Sci USA 100:8721–8726

Gosalia DN, Salisbury CM, Ellman JA, Diamond SL (2005) High throughput substrate specificity profiling of serine and cysteine proteases using solution-phase fluorogenic peptide microarrays. Mol Cell Proteomics 4:626–636

Guo JP, Petric M, Campbell W, McGeer PL (2004) SARS corona virus peptides recognized by antibodies in the sera of convalescent cases. Virology 324:251–256

Heneweer M, Houtman R, Poortman J, Groot M, Maliepaard C, Peijnenburg A (2007) Estrogenic effects in the immature rat uterus after dietary exposure to ethinylestradiol and zearalenone using a systems biology approach. Toxicol Sci 99:303–314

Hilhorst R, Houkes L, van den Berg A, Ruijtenbeek R (2009) Peptide microarrays for detailed, high-throughput substrate identification, kinetic characterization, and inhibition studies on protein kinase A. Anal Biochem 387:150–161

Hilpert K, Winkler DF, Hancock RE (2007) Peptide arrays on cellulose support: SPOT synthesis, a time and cost efficient method for synthesis of large numbers of peptides in a parallel and addressable fashion. Nat Protoc 2:1333–1349

Huang H, Li L, Wu C, Schibli D, Colwill K, Ma S, Li C, Roy P, Ho K, Songyang Z, Pawson T, Gao Y, Li SS (2008) Defining the specificity space of the human SRC homology 2 domain. Mol Cell Proteomics 7:768–784

Inamori K, Kyo M, Matsukawa K, Inoue Y, Sonoda T, Tatematsu K, Tanizawa K, Mori T, Katayama Y (2008) Optimal surface chemistry for peptide immobilization in on-chip phosphorylation analysis. Anal Chem 80:643–650

Jalal S, Arsenault R, Potter AA, Babiuk LA, Griebel PJ, Napper S (2009) Genome to kinome: species-specific peptide arrays for kinome analysis. Sci Signal 2:1–10

Köhn M, Gutierrez-Rodriguez M, Jonkheijm P, Wetzel S, Wacker R, Schroeder H, Prinz H, Niemeyer CM, Breinbauer R, Szedlacsek SE, Waldmann H (2007) A microarray strategy for mapping the substrate specificity of protein tyrosine phosphatase. Angew Chem Int Ed Engl 46:7700–7703

Koppen A, Houtman R, Pijnenburg D, Jeninga EH, Ruijtenbeek R, Kalkhoven E (2009) Nuclear receptor-coregulator interaction profiling identifies TRIP3 as a novel peroxisome proliferator-activated receptor gamma cofactor. Mol Cell Proteomics 8:2212–2226

Kwak EL, Bang YJ, Camidge DR, Shaw AT, Solomon B, Maki RG, Ou SH, Dezube BJ, Jänne PA, Costa DB, Varella-Garcia M, Kim WH, Lynch TJ, Fidias P, Stubbs H, Engelman JA, Sequist LV, Tan W, Gandhi L, Mino-Kenudson M, Wei GC, Shreeve SM, Ratain MJ, Settleman J, Christensen JG, Haber DA, Wilner K, Salgia R, Shapiro GI, Clark JW, Iafrate AJ (2010) Anaplastic lymphoma kinase inhibition in non-small-cell lung cancer. N Engl J Med 363:1693–1703

Lemeer S, Jopling C, Naji F, Ruijtenbeek R, Slijper M, Heck AJ, den Hertog J (2007) Protein-tyrosine kinase activity profiling in knock down zebrafish embryos. PLoS One 2(7):e581

Lesaicherre ML, Uttamchandani M, Chen GY, Yao SQ (2002) Developing site-specific immobilization strategies of peptides in a microarray. Bioorg Med Chem Lett 12:2079–2083

Lipshutz RJ, Fodor SP, Gingeras TR, Lockhart DJ (1999) High density synthetic oligonucleotide arrays. Nature Genetics 21:20–24

Luo K, Zhou P, Lodish HF (1995) The specificity of the transforming growth factor beta receptor kinases determined by a spatially addressable peptide library. Proc Natl Acad Sci USA 92:11761–11765

Mahrenholz CC, Tapia V, Stigler RD, Volkmer R (2010) A study to assess the cross-reactivity of cellulose membrane-bound peptides with detection systems: an analysis at the amino acid level. J Pept Sci 16:297–302

McGall GH, Fidanza JA (2001) Photolithographic synthesis of high-density oligonucleotide arrays. Methods Mol Biol 170:71–101

Merrifield RB (1965a) Automated synthesis of peptides. Science 150:178–185

Merrifield RB (1965b) Solid-phase peptide syntheses. Endeavour 24:3–7

Moelans CB, de Weger RA, Van der Wall E, van Diest PJ (2011) Current technologies for HER2 testing in breast cancer. Crit Rev Oncol Hematol 80:380–392

Neuman de Vegvar HE, Amara RR, Steinman L, Utz PJ, Robinson HL, Robinson WH (2003) Microarray profiling of antibody responses against simian-human immunodeficiency virus: postchallenge convergence of reactivities independent of host histocompatibility type and vaccine regimen. J Virol 77:11125–11138

Pawson T (2007) Dynamic control of signaling by modular adaptor proteins. Curr Opin Cell Biol 19:112–116

Pease AC, Solas D, Sullivan EJ, Cronin MT, Holmes CP, Fodor SP (1994) Light-generated oligonucleotide arrays for rapid DNA sequence analysis. Proc Natl Acad Sci USA 91:5022–5026

Piersma SR, Labots M, Verheul HM, Jiménez CR (2010) Strategies for kinome profiling in cancer and potential clinical applications: chemical proteomics and array-based methods. Anal Bioanal Chem 397:3163–3171

Poot AJ, van Ameijde J, Slijper M, van den Berg A, Hilhorst R, Ruijtenbeek R, Rijkers DT, Liskamp RM (2009) Development of selective bisubstrate-based inhibitors against protein kinase C (PKC) isozymes by using dynamic peptide microarrays. Chembiochem 10:2042–2051

Hilhorst R, Schaake E, van Pel R, Nederlof P M, Houkes L, Mommersteeg M, de Wijn R, van den Heuvel M, Baas P, Ruijtenbeek R, Klomp HM (2011) Blind prediction of response to erlotinib in early-stage non-small cell lung cancer (NSCLC) in a neoadjuvant setting based on kinase activity profiles. Annual meeting of the American Society of Clinical Oncology. ASCO #10521

Salisbury CM, Maly DJ, Ellman JA (2002) Peptide microarrays for the determination of protease substrate specificity. J Am Chem Soc 124:14868–14870

Schrage YM, Briaire-de Bruijn IH, de Miranda NF, van Oosterwijk J, Taminiau AH, van Wezel T, Hogendoorn PC, Bovée JV (2009) Kinome profiling of chondrosarcoma reveals SRC-pathway activity and dasatinib as option for treatment. Cancer Res 69:6216–6222

Schutkowski M, Reimer U, Panse S, Dong L, Lizcano JM, Alessi DR (2004) High-content peptide microarrays for deciphering kinase specificity and biology. Angew Chem 116:2725–2728

Schutkowski M, Reineke U, Reimer U (2005) Peptide arrays for kinase profiling. Chembiochem 6:513–521

Shin DS, Lee KN, Yoo BW, Kim J, Kim M, Kim YK, Lee YS (2010) Automated maskless photolithography system for peptide microarray synthesis on a chip. J Comb Chem 12:463–471

Singh-Gasson S, Green RD, Yue Y, Nelson C, Blattner F, Sussman MR, Cerrina F (1999) Maskless fabrication of light-directed oligonucleotide microarrays using a digital micromirror array. Nat Biotechnol 17:974–978

Sinzinger MD, Brock R (2010) Peptide microarrays for a network analysis of changes in molecular interactions in cellular signalling. In: Systems biology for signaling networks, vol 1, System biology. Springer, New York, pp 703–718

Stadler V, Beyer M, König K, Nesterov A, Torralba G, Lindenstruth V, Hausmann M, Bischoff FR, Breitling F (2007) Multifunctional CMOS microchip coatings for protein and peptide arrays. J Proteome Res 6:3197–3202

Stadler V, Felgenhauer T, Beyer M, Fernandez S, Leibe K, Guettler S, Groening M, Torralba G, Lindenstruth V, Nesterov A, Block I, Pipkorn R, Poustka A, Bischoff FR, Breitling F (2008)

Angew Chem 120:7241–7244; (2008)Combinatorial Synthesis of Peptide Arrays with a Laser Printer. Angew Chem Int Ed 47:7132–7135

Stoevesandt O, Elbs M, Köhler K, Lellouch AC, Fischer R, André T, Brock R (2005) Peptide microarrays for the detection of molecular interactions in cellular signal transduction. Proteomics 5:2010–2017

Sun H, Lu CH, Uttamchandani M, Xia Y, Liou YC, Yao SQ (2008) Peptide microarray for high-throughput determination of phosphatase specificity and biology. Angew Chem Int Ed Engl 47:1698–1702

Sun H, Tan LP, Gao L, Yao SQ (2009) High-throughput screening of catalytically inactive mutants of protein tyrosine phosphatases (PTPs) in a phosphopeptide microarray. Chem Commun 14:677–679

Thiele A, Weiwad M, Zerweck J, Fischer G, Schutkowski M (2010) High density peptide microarrays for proteome-wide fingerprinting of kinase activities in cell lysates. Methods Mol Biol 669:173–181

Thiele A, Stangl GI, Schutkowski M (2011) Deciphering enzyme function using peptide arrays. Mol Biotechnol 49:283–305. doi:10.1007/s12033-011-9402-x

Thilakarathne PJ, Clement L, Lin D, Shkedy Z, Kasim A, Talloen W, Versele M, Verbeke G (2011) The use of semi-parametric mixed models to analyze PamChip peptide array data: an application to an oncology experiment. Bioinformatics 27:2859–2865

Tonikian R, Zhang Y, Boone C, Sidhu SS (2008) Identifying specificity profiles for peptide recognition modules from phage-displayed peptide libraries. Nat Protoc 2:1368–1386

Tonikian R, Xin X, Toret CP, Gfeller D, Landgraf C, Panni S, Paoluzi S, Castagnoli L, Currell B, Seshagiri S, Yu H, Winsor B, Vidal M, Gerstein MB, Bader GD, Volkmer R, Cesareni G, Drubin DG, Kim PM, Sidhu SS, Boone C (2009) Bayesian modeling of the yeast SH3 domain interactome predicts spatiotemporal dynamics of endocytosis proteins. PLoS Biol 7:e1000218

Tuynman JB, Vermeulen L, Boon EM, Kemper K, Zwinderman AH, Peppelenbosch MP, Richel DJ (2008) Cyclooxygenase-2 inhibition inhibits c-Met kinase activity and Wnt activity in colon cancer. Cancer Res 68:1213–1220

van Ameijde J, Poot AJ, van Wandelen LT, Wammes AE, Ruijtenbeek R, Rijkers DT, Liskamp RM (2010) Preparation of novel alkylated arginine derivatives suitable for click-cycloaddition chemistry and their incorporation into pseudosubstrate- and bisubstrate-based kinase inhibitors. Org Biomol Chem 8:1629–1639

Versele M, Talloen W, Rockx C, Geerts T, Janssen B, Lavrijssen T, King P, Göhlmann HW, Page M, Perera T (2009) Response prediction to a multitargeted kinase inhibitor in cancer cell lines and xenograft tumors using high-content tyrosine peptide arrays with a kinetic readout. Mol Cancer Ther 8:1846–1855

Virshup DM, Shenolikar S (2009) From promiscuity to precision: protein phosphatases get a makeover. Mol Cell 33:537–545

Wang P, Fu H, Snavley DF, Freitas MA, Pei D (2002) Screening combinatorial libraries by mass spectrometry. 2. Identification of optimal substrates of protein tyrosine phosphatase SHP-1. Biochemistry 41:6202–6210

Winssinger N, Damoiseaux R, Tully DC, Geierstanger BH, Burdick K, Harris JL (2004) PNA-encoded protease substrate microarrays. Chem Biol 11:1351–1360

Zander N, Beutling U, Dikmans A, Thiele S, Frank R (2005) A special cellulose membrane support for the combinatorial and parallel synthesis of peptide libraries suitable for the SC2-type manufacturing of high density multi-purpose chemical micro-arrays. In: Flegel M. et al. (eds) Peptides 2004: proceedings of the third international and 28th European peptide symposium, pp 405–406

Zhang ZY, Maclean D, McNamara DJ, Sawyer TK, Dixon JE (1994) Protein tyrosine phosphatase substrate specificity: size and phosphotyrosine positioning requirements in peptide substrates. Biochemistry 33:2285–2290

Zhao S, Lee EY (1997) A protein phosphatase-1-binding motif identified by the panning of a random peptide display library. J Biol Chem 272:28368–28372

Chapter 8
Slow Off-Rate Modified Aptamer Arrays for Biomarker Discovery and Diagnostic Applications

Jeffrey J. Walker, Edward N. Brody, and Larry Gold

8.1 Introduction

The fields of "omics" have fundamentally influenced how we do research, how we understand biological systems, and how we search for biomarkers for new diagnostics. Omics is increasingly appreciated for generating hypotheses by enabling unbiased discovery of new phenomena to guide further experiments (Brody et al. 2010; Glass 2010; Kell and Oliver 2004; Lander 2010; Nabel 2009).

In 1997, we concluded that, of all the omics, proteomics would most likely yield the most valuable biological and health information, and we set out to use aptamers (single-stranded DNA molecules with high affinity for specific proteins) to conduct unbiased proteomic measurements to interrogate biological systems and identify biomarkers of human disease in order to advance personalized medicine with new diagnostics and treatments. Our goal was to transform proteomics in the way that DNA microarrays transformed genomics. Over the past few years, we have made great progress, and microarrays have played an important role. For highly multiplexed measurements, microarrays elegantly solve the problem of multiplexing through spatial addressing. Advancing microarray technology has been key to advances in omics sciences, and today a large variety of technologies enable applications, from inexpensive microarrays for tens of DNA targets that enable the latest crop of molecular diagnostics (i.e., nucleic acid-based diagnostics) to high-performance microarrays with millions of features for quantitative transcriptomics.

J.J. Walker (✉) • E.N. Brody
SomaLogic, 2945 Wilderness Place, Boulder, CO 80301, USA
e-mail: jwalker@somalogic.com

L. Gold
SomaLogic, 2945 Wilderness Place, Boulder, CO 80301, USA

Department of Molecular, Cellular, and Developmental Biology, University of Colorado, Boulder, CO 80309, USA

B. Jordan (ed.), *Microarrays in Diagnostics and Biomarker Development*,
DOI 10.1007/978-3-642-28203-4_8, © Springer-Verlag Berlin Heidelberg 2012

We developed a high-throughput, high-content proteomics platform (SOMAscan), allowing unbiased discovery of protein biomarkers (Gold et al. 2010a). The current SOMAscan discovery assay measures over 1,000 human proteins simultaneously in a small sample volume (e.g., ~10 μL of serum, plasma, tissue homogenate) with low limits of quantification (median LLOQ 300 fM), broad dynamic range [mean 4 logs per analyte, >7 logs overall in plasma or serum (~30 fM–1 μM, using sample dilution)], and low CVs (<5%) (Gold et al. 2010a). The assay in its current configuration has a throughput of hundreds of samples per day on one standard apparatus, and the throughput is easily scaled. Thus, we have made proteomic measurements as fast and easy as nucleic acid-based omics (genomics, transcriptomics, etc.) and enabled unbiased interrogation of a large part of the proteome for biomarker discovery.

To date, we have run over 24,000 analyses of clinical samples, mostly serum and plasma, from a clinical biobank we built with collaborators over the past decade to study human diseases ranging from cancer to depression. From these clinical studies, we have identified numerous potential biomarkers for many diseases. The results of these studies are feeding a diagnostics development pipeline that includes several tests in development with a large commercial partner.

Here we will discuss the development of our proteomics technology, based on slow off-rate modified aptamer (SOMAmer) arrays, and its application to developing commercial diagnostics. Successful commercial development relies on navigating and executing a complex process that starts with developing a product concept [intended use, clinical benefit, health economic analysis, payment strategy (e.g., reimbursement)], a clinical study design (target population, appropriate control population, study type, sample size power analysis, validation strategy), collecting and processing appropriate biospecimens, conducting proteomic analysis, identifying biomarkers, developing a diagnostic algorithm and test, transferring the test to the intended clinical platform, developing and manufacturing test kits, validating the test in independent clinical studies, obtaining regulatory approval and payment mechanism (e.g., CPT reimbursement code), and finally ending with the launch of the product into the commercial marketplace.

8.2 The Challenge of Proteomics

The human proteome contains an estimated 20,000 proteins that span a concentration range of at least 12 logs (~mM–pM) and probably more (Anderson 2002). Proteomic measurements demand extreme sensitivity, specificity, dynamic range, and accurate quantification (Eaton et al. 1995; Zichi et al. 2008). Furthermore, proteins are more difficult to measure than DNA. Proteins have no equivalent to the complementarity that allows specific nucleic acid hybridization.

Early efforts to conduct high-scale proteomics began with two-dimensional gels, pioneered by Pat O'Farrell in his extraordinary work as a graduate student (O'Farrell 1975). The quality of these first two-dimensional gels was remarkable;

approximately 1,100 proteins were resolved quantitatively over about 3 logs of concentration. More recently, two technologies have dominated proteomics, but they are, so far, inadequate for deep unbiased proteomic measurements—antibody-based affinity assays and mass spectrometry.

The conclusions that we draw from a large body of work are that large antibody microarrays, in either single antibody or sandwich format (e.g., ELISAs), do not scale easily to large numbers of analytes, as discussed in Chap. 6 [also see Brody et al. (2010); Zichi et al. (2008)]. Furthermore, despite great advances and promise for mass spectrometry in clinical proteomics, many challenges remain, including issues of sensitivity (typically just under nM in current approaches), specificity, reproducibility, throughput, and cost (Addona et al. 2009; Aebersold 2009; Bell et al. 2009; Liotta and Petricoin 2010; Mitchell 2010; Pan et al. 2009; Service 2008; Silberring and Ciborowski 2010). Nevertheless, we acknowledge that every MS parameter will improve over time.

8.3 A New Approach to Proteomics

Armed with an understanding of molecular binding and binding kinetics and how these attributes affected antibody assay performance, and with a powerful technology to create and fine-tune fully synthetic affinity molecules by selection from combinatorial libraries (below), we set out to do whatever it took to develop a robust proteomics platform. After several iterations, we achieved this goal and applied it to biomarker discovery and developing commercial diagnostics. We have learned a great deal in both developing the proteomics technology and applying it to commercial diagnostics development, which we share here. Today we employ microarrays every day to search for protein biomarkers of human health and disease and to develop these discoveries into commercial diagnostic products.

8.3.1 Aptamers, Photoaptamers, and SOMAmers

Aptamers are a class of molecules that were described in two key papers published in 1990 (Ellington and Szostak 1990; Tuerk and Gold 1990) and are analogous to monoclonal antibodies in terms of protein binding, but they are made of single-stranded nucleic acids (RNA or DNA traditionally). Aptamers are selected from large combinatorial libraries of randomly synthesized oligonucleotides by the process of *s*ystematic *e*nrichment of *l*igands by *ex*ponential enrichment (SELEX) (Tuerk and Gold 1990). Aptamers usually bind native protein conformations in contrast to antibodies, which often bind linear epitopes. Aptamers are an alternative class of reagents to use for highly multiplexed protein measurements (Bock et al. 2004; Gold et al. 2010a; Keeney et al. 2009; Ostroff et al. 2010a, 2010b; Smith et al. 2003; Tuerk and Gold 1990; Zichi et al. 2002, 2008). About 10 years ago, the entire literature for aptamers included the identification of classic aptamers for only approximately 100 proteins. At that time, we set a goal to generate aptamers good enough to measure proteins in complex matrices for a large number of human proteins and an assay platform to allow ELISA-like performance, but without sandwiches (Gold et al. 2010a, 2010b; Zichi et al. 2008).

The problems intrinsic to the use of a single binding reagent to quantify a human protein in plasma or serum, for example, have been described (Rifai et al. 2006; Zichi et al. 2008). In essence, measuring a plasma protein with a single reagent against a protein background that spans 12 logs of concentration, and relying on equilibrium binding alone, is not physically possible (Gold et al. 2010a; Zichi et al. 2008). Therefore, we had to figure out another parameter we could use to provide a second element of specificity beyond low equilibrium constants (K_ds below 1 nM, for example) using only one protein-binding reagent (Bock et al. 2004; Petach 2002; Smith 2002). The sandwich assays (e.g., ELISA) solve this problem with two protein-binding reagents, each of which provides a separate element of specificity (Brody et al. 2010).

The path to the final solution was illuminated in part by the papers of John Hopfield (Hopfield 1974; Hopfield et al. 1976), which lead us to develop a new class of molecules we call slow off-rate modified aptamers (SOMAmers). Hopfield recognized that the accuracy of protein synthesis by the ribosome could not be explained by equilibrium binding of amino-acyl tRNAs alone and correctly deduced a mechanism, dubbed kinetic proofreading, that divides the event into two separate binding reactions, each with its own element of specificity, separated by GTP hydrolysis.

We first tried photoaptamers, which provided a photoactivated cross-linking step conceptually equivalent to Hopfield's post-GTP hydrolysis kinetic proofreading step. Photoaptamers are selected with a novel in vitro selection methodology called photochemical systematic evolution of 'ligands by exponential enrichment (PhotoSELEX) (Golden et al. 2000).

We started down the proteomics path with microarrays printed with photoaptamers—aptamers with the ability to form covalent bonds to their target proteins when activated with light (Zichi et al. 2002). Photoaptamers were developed as affinity reagents specifically for use with microarrays, and the underlying idea was that once cognate complexes were covalently bound, extreme washing could solve the issue of nonspecific binding to microarray surfaces. We had success with a moderate number of targets using a universal protein stain for detection (Zichi et al. 2002) as well as antibodies for additional specificity and detection (Bock et al. 2004).

After much experimentation, we developed a more robust approach to multiplexing in which SOMAmers were bound to complex matrices in solution, and simple biochemistry allowed us to quantify the bound proteins by measuring the bound SOMAmers. We discovered that we could use equilibrium binding as the first element of specificity, followed by kinetic selection for stable complexes between aptamers and target proteins with slow dissociation rates as the required second element of specificity. In order to build our proteomics platform, we invented a new class of aptamers that we call SOMAmers (for *s*low *o*ff-rate *m*odified *a*pta*mers*). The idea was to generate novel reagents that had the low K_ds

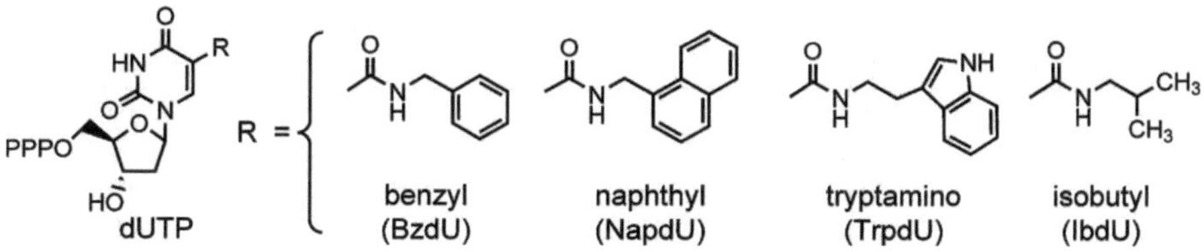

Fig. 8.1 Modified nucleotides. Nucleotide triphosphate analogs modified at the 5-position (R) of uridine (dUTP): 5-benzylaminocarbonyl-dU (BndU); 5-naphthylmethylaminocarbonyl-dU (NapdU): 5-tryptaminocarbonyl-dU (TrpdU); and 5-isobutylaminocarbonyl-dU (iBudU)

of classic aptamers and that in addition had very slow dissociation rates with their cognate analytes but fast dissociation rates for most other proteins in plasma or serum. Dissociation rates for most aptamers with plasma proteins are fast, with half-lives of a few seconds. We reasoned that we simply had to select for slow dissociation rates in the SELEX process itself. Simultaneously, we realized that novel nucleotide adducts, with new functionalities at the five positions on pyrimidines, as shown in Fig. 8.1 (Vaught et al. 2004, 2010), would provide chemistry missing in classic aptamers and might yield SOMAmers that bound their cognate partners through extra binding interactions (Gold et al. 2010a; Zichi et al. 2008). The on-rates are typically slower than diffusion limited, but the off-rates are also slow (half-lives >30 min) (Gold et al. 2010a).

The critical innovation of SOMAmers substantially enhances their binding properties and allows the selection of high-quality SOMAmers to nearly any target protein. This is illustrated in Table 8.1, which shows the results of selections using modified and unmodified nucleotides targeting 13 human proteins that had repeatedly failed SELEX with unmodified DNA. We used GA733-1 protein as a positive control because it had previously yielded high-affinity aptamers with unmodified DNA SELEX. The results demonstrate that only SELEX for SOMAmers (containing modified nucleotides) and not aptamers (consisting of only standard, unmodified DNA nucleotides) yielded high-affinity aptamers to these difficult proteins. Different modifications worked better with different proteins in SELEX, which shows that applying multiple modifications to the same target increases the probability of success.

We have determined the structure of three SOMAmers complexed with their protein targets at between 2.2 and 2.5 Å resolution by X-ray crystallography (unpublished), and more structures are under way. As predicted, contacts between each protein and SOMAmer are extensive and contain elements never observed previously in either classic aptamer–protein structures (Huang et al. 2003; Long et al. 2008) or structures between natural RNAs and their binding partners (Vicens and Cech 2006). In one of the solved structures, the SOMAmer was identified from a single-stranded DNA library where 5-benzyl-dUMP (Vaught et al. 2010) was substituted for every "T." We found interactions between benzyl groups and amino acids, benzyl groups stacking on other nucleotides, and a remarkable compact hydrophobic "turn" with benzyl groups at its core. Small modifications to standard nucleotide chemistry open a new world of possible structures; although SOMAmers

Table 8.1 SELEX library affinities [K_d, (M)] with unmodified and modified nucleotides

Target protein	dT	Benzyl-dU	Isobutyl-dU	Tryptamino-dU
4-1BB[a]	Failed	6×10^{-9}	Failed	4×10^{-9}
B7[a]	Failed	1×10^{-8}	Failed	7×10^{-9}
B7-2[a]	Failed	Failed	Failed	6×10^{-9}
CTLA-4[a]	Failed	Failed	Failed	1×10^{-9}
sE-Selectin[a]	Failed	Failed	Failed	2×10^{-9}
Fractalkine/CXC3L-1	Failed	Failed	Failed	5×10^{-11}
GA733-1 protein[a]	9×10^{-9}	3×10^{-9}	5×10^{-9}	5×10^{-10}
gp130, soluble[a]	Failed	6×10^{-9}	2×10^{-8}	1×10^{-9}
HMG-1	Failed	Failed	2×10^{-8}	5×10^{-9}
IR	Failed	2×10^{-9}	1×10^{-8}	2×10^{-10}
Osteoprotegrin[a]	Failed	5×10^{-9}	9×10^{-9}	2×10^{-10}
PAI-1	Failed	4×10^{-10}	9×10^{-10}	2×10^{-10}
P-cadherin[a]	Failed	4×10^{-9}	5×10^{-9}	3×10^{-9}
sLeptin R[a]	Failed	2×10^{-9}	Failed	5×10^{-10}

[a]The protein used was expressed as a fusion to the Fx of human IgG_1. No detectable binding of the active library to an alternate Fc fusion protein was observed. Table reproduced from (Gold et al. 2010a)

are aptamers obtained through a modified SELEX process, their properties are completely unique. Modification of the pyrimidines of oligonucleotide libraries and understanding the high-resolution atomic structures is part of our continuing efforts to improve SOMAmers.

8.3.2 Solution Kinetics and DNA Microarrays: A Critical Transformation

The core idea of the SOMAscan proteomics assay is to selectively bind many SOMAmers to their target proteins in solution, to purify specific (cognate) from nonspecific SOMAmer–protein complexes, and to release and ultimately quantify the SOMAmers. The key purification step was achieved with our version of kinetic proofreading, where nonspecific complexes dissociate rapidly compared to specific complexes, and a polyanionic competitor (e.g., dextran sulfate) prevents reformation of new SOMAmer–protein complexes. In the end, a DNA measurement yields proteomic quantification because SOMAmers themselves are DNA molecules. The assay is described in Gold et al. (2010a).

The SOMAscan assay, shown in Fig. 8.2, takes advantage of the dual nature of SOMAmers as both folded protein-binding elements with defined shapes and unique nucleotide sequences recognizable by complementary hybridization probes. Briefly, the sample is incubated with a mixture of SOMAmers (each containing a biotin, a photocleavable group, and a fluorescent tag) followed by capture of all SOMAmer–protein complexes on streptavidin beads (Catch-1) (Fig. 8.2a–c). After stringent washing of the beads to remove unbound proteins, bead-associated

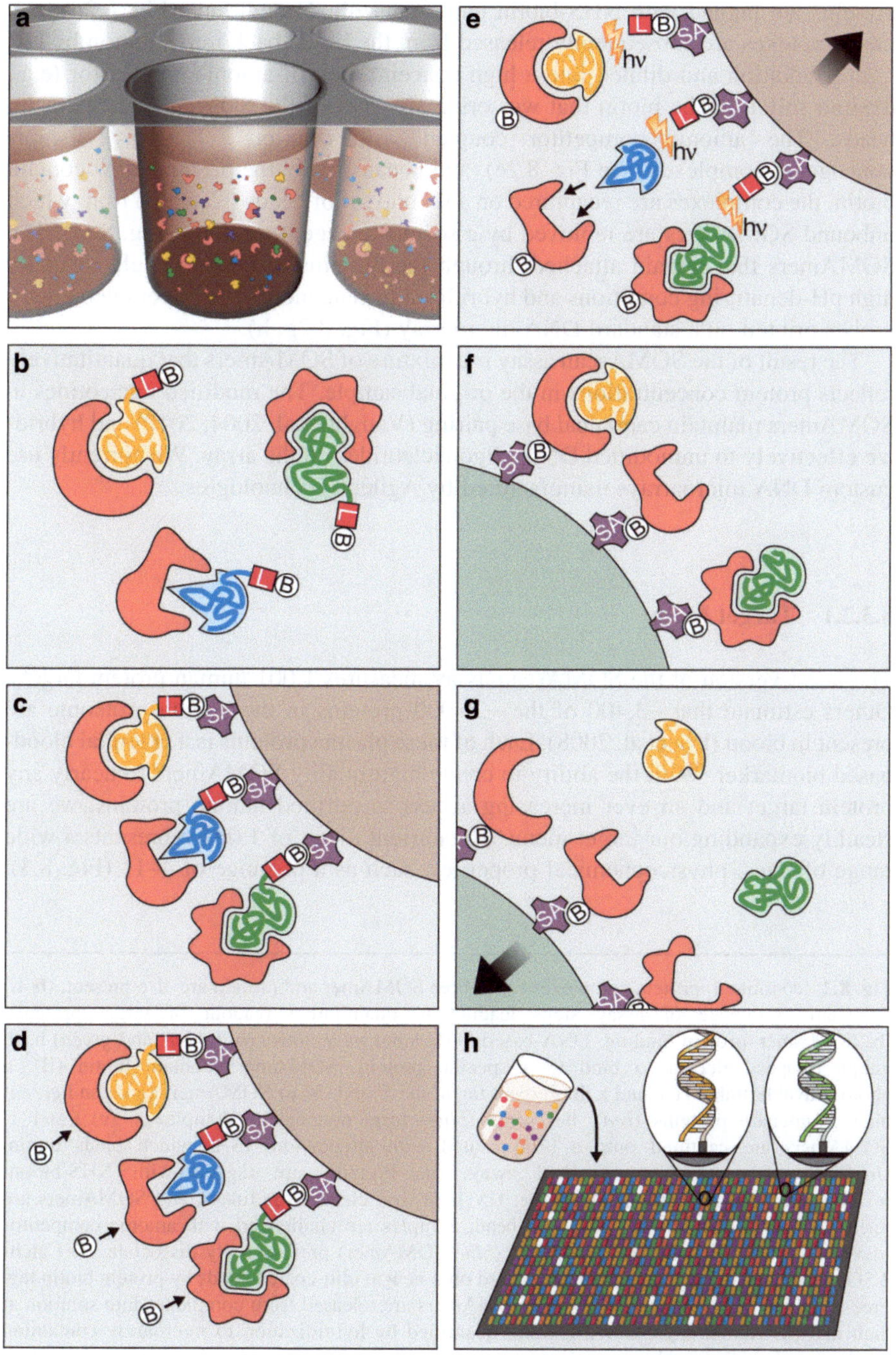

Fig. 8.2 Principle of multiplex SOMAmer affinity assay. (**a**) Binding. SOMAmers and samples are mixed in 96-well microwell plates and allowed to bind. Cognate and noncognate SOMAmer-target

proteins are tagged with NHS-biotin under controlled conditions (Fig. 8.2d), and the complexes are subsequently released from the beads back into solution by UV light irradiation and diluted into a high concentration of anionic competitor (e.g., dextran sulfate). The biotin that was originally part of the SOMAmer remains on beads. The anionic competitor coupled with dilution selectively disrupts noncognate complexes (see Fig. 8.2e), and because only the proteins now contain biotin, the complexes are recaptured on a second set of beads (Catch-2) from which unbound SOMAmers are removed by a second stringent washing (Fig. 8.2f). The SOMAmers that remain attached through the proteins to beads are eluted under high pH-denaturing conditions and hybridized to sequence-specific complementary probes printed on a standard DNA microarray (Fig. 8.2g, h).

The result of the SOMAscan assay is a mixture of SOMAmers that quantitatively reflects protein concentrations in the original sample. The modified nucleotides in SOMAmers maintain canonical base pairing (Vaught et al. 2004, 2010) and hybridize effectively to unmodified DNA oligonucleotides on the array. We currently use custom DNA microarrays manufactured by Agilent Technologies.

8.3.2.1 Target Menu

The latest version of the SOMAscan assay measures 1,001 human protein targets. Others estimate that ~3,400 of the ~20,000 proteins in the human proteome are present in blood (Lin et al. 2008). Each of these plasma proteins is a potential blood-based biomarker. With the ability to create high-quality SOMAmers to nearly any protein target and an ever increasing access to purified human proteins, we are steadily expanding our target menu. The current menu of 1,001 represents a wide range of sizes, physicochemical properties, such as a p*I* range of 4–11 (Fig. 8.3),

Fig. 8.2 (continued) protein complexes form. Free SOMAmer and protein are also present. (**b–h**) Schematic sequence of assay steps leading to quantitative readout of target proteins. (**b**) SOMAmer–protein binding: DNA-based SOMAmer molecules (*gold, blue,* and *green*) have unique shapes selected to bind to a specific protein. SOMAmers contain biotin (B), a photocleavable linker (L), and a fluorescent tag at the 5′ end. Most SOMAmers (*gold* and *green*) bind to cognate proteins (*red*), but some (*blue*) form noncognate complexes. (**c**) Catch-1. SOMAmers are captured onto a bead coated with streptavidin (SA) which binds biotin. Uncomplexed proteins are washed away. (**d**) Proteins are tagged with NHS-biotin. (**e**) Photocleavage and kinetic challenge. UV light (hv) cleaves the linker, and SOMAmers are released from beads, leaving biotin on bead. Samples are challenged with anionic competitor (dextran sulfate). Noncognate complexes (*blue* SOMAmer) preferentially dissociate. (**f**) Catch-2 SOMAmer–protein complexes are captured onto new avidin-coated beads by protein biotin tag. Free SOMAmers are washed away. (**g**) SOMAmers are released from complexes into solution at high pH. (**h**) Remaining SOMAmers are quantified by hybridization to microarray containing single-stranded DNA probes complementary to SOMAmer DNA sequence, which form a double-stranded helix. Hybridized SOMAmers are detected by fluorescent tags when the array is scanned. Figure reproduced form Gold et al. (2010a)

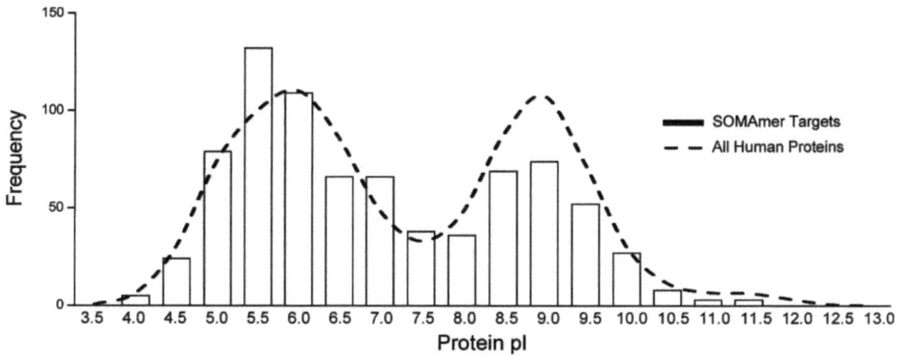

Fig. 8.3 Target isoelectric points. Distribution of isoelectric points (p*I*) of proteins for which SOMAmers have been selected (*bars*) and of all human protein chains in UniProt (*dashed line*). Data shown for 813 proteins, reproduced from Gold et al. (2010a)

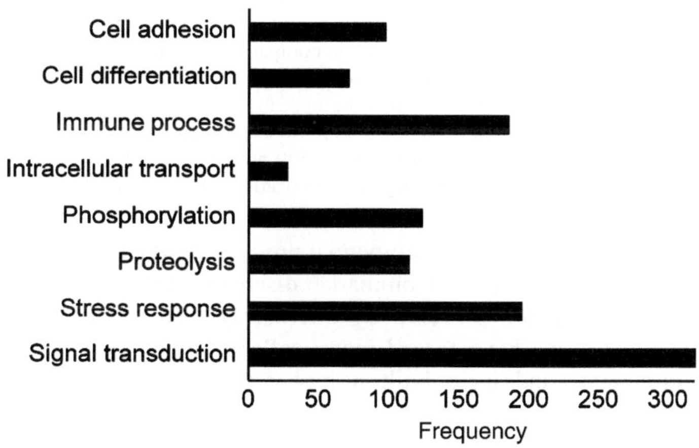

Fig. 8.4 Protein target menu gene ontology. Distribution of most common gene ontology terms associated with the proteins measured by the current array. Data shown for 813 proteins, reproduced from Gold et al. (2010a)

and biological functions from a range of molecular pathways and gene families (Fig. 8.4).

8.3.2.2 Diagnostics

We are applying the SOMAscan proteomics platform to identify biomarkers and develop diagnostics for many human disease areas, including oncology, cardiovascular, renal, neurological, inflammatory, infectious, as well as for health and wellness. Promising results from a number of these studies are summarized in Fig. 8.5. These results are the foundation of our commercial diagnostics pipeline,

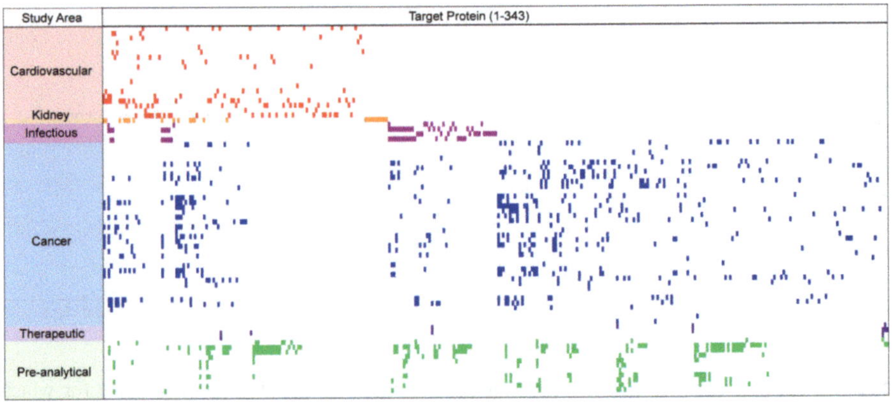

Fig. 8.5 Preliminary proteomics biomarker discovery map. Each row represents one of 77 statistical comparisons of clinical serum or plasma samples (disease and matched controls) that address specific clinical questions in the study area indicated. Each column represents a unique protein measured with the SOMAscan proteomics discovery platform. A *colored rectangle* represents a potential biomarker for the respective comparison that met the following criteria in the statistical comparison: Kolmogorov–Smirnov (KS) test distance >0.3, KS test p-value <0.01, KS distance rank in top 25 (highest), and p-value rank in top 25 (lowest). Ties in ranking were allowed, so >25 preliminary biomarkers are possible for each comparison. The map shows 334 of ~850 possible protein targets measured. The other ~500 proteins were not identified as potential biomarkers in the 77 experiments. Figure reproduced from Brody et al. (2010)

which is fed continuously by our commercial product development cycle. The core of this process is high-scale protein biomarker discovery enabled by the SOMAscan platform, which is in turn powered by microarrays. To support our efforts to develop diagnostics, we have forged many collaborations, including with many academic clinicians, and have built a biobank with >12,000 clinical blood specimens.

8.4 Diagnostics Development Process

Developing commercial diagnostics is a challenging discipline that demands rigorous planning and execution from the earliest stages of development. Great enthusiasm for biomarkers and a new generation of diagnostics accompanied the sequencing of the human genome a decade ago. This led to a surge of published biomarker discoveries as basic scientists and clinical scientists dove into each other's relatively unfamiliar waters. This is true for both protein-based biomarkers and DNA-based biomarkers. What has happened since is that nearly all initial biomarker discoveries have failed to validate in subsequent clinical studies (Rifai et al. 2006).

Thus, initial enthusiasm, fueled by great advances in technology and exponential growth in data generation, quickly caught up with the hard realities of clinical research, validation, and the translation to diagnostic products. In fact, a recent survey found a median of just one protein biomarker validated per year for the past 15 years (Anderson 2010). Much has been written about the reasons why so few biomarkers are validated (Anderson 2010; Ostroff et al. 2010a; Rifai et al. 2006), which include technical issues with proteomic measurements (mentioned above), preanalytical variability (e.g., sample handling) (Ostroff et al. 2010a), as well as study design, statistical analysis (e.g., overfitting), and validation (Ostroff et al. 2010b; Pepe and Feng 2011; Pepe et al. 2008; Ransohoff 2004; Ransohoff and Gourlay 2010).

The silver lining is that the field is emerging stronger and smarter than ever with a sophisticated understanding at all levels from molecular technology and clinical study design to validation and regulatory requirement. The field has faced steep technological and methodological learning curves, but we remain as enthusiastic as ever about the promise of protein biomarkers and a new generation of diagnostics to transform medicine.

Here we outline in relatively broad strokes our diagnostic product development process, and below we review two examples diagnostics we are currently developing into products: lung cancer and chronic kidney disease (CKD).

8.4.1 Product Concept

This critical first step defines the intended use of the product and characterizes the weight of evidence, the nature of evidence, and the minimum performance required to qualify the product for its intended use. It is essential to have an accurate diagnosis as a "truth standard," such as a malignancy confirmed by biopsy. It can also be useful to have an alternative test to use as a comparative standard. Important commercial considerations include market research and health economic analysis to forecast market size and price.

8.4.2 Study Design

Much has been written about approaches to study design for discovery and validation of biomarkers, and some key papers (Pepe et al. 2001, 2008; Pepe and Feng 2011; Ransohoff 2004; Ransohoff and Gourlay 2010) provide an excellent introduction to this essential topic. Basic considerations include a study design that accurately reflects the intended use of the product concept, an appropriately powered study for the intended statistical analysis, and a sound approach for verification and ultimately independent validation of the results. It is good practice to predefine the study plan, statistical analysis plan, and minimum performance criteria for acceptance or validation. It is also wise to begin thinking here in the first step about the last steps in development, seeking regulatory approval for the diagnostic and payment mechanisms, especially in the case of systems with third-party payers.

8.4.3 Sample Collection and Provenance

A big lesson learned in proteomic biomarker discovery is that samples, particularly blood samples, are subject to preanalytical variability that can confound studies and lead to false discovery. Variability is known to arise from variations in procedures for sample collection, processing, storage, transport, and handling (Ostroff et al. 2010a). Sample provenance and documentation are equally important. Therefore, it is critical to properly control or assess sample quality, provenance, and documentation because biomarker discovery relies on analysis of the sample and its associated information. In the course of running many biomarker discovery and control studies, we have learned a great deal about preanalytical variability (Ostroff et al. 2010a). Consequently, we developed in vitro and in silico methods to prevent, detect, and even compensate for preanalytical variability (unpublished).

8.4.4 Biomarker Identification

Candidate biomarkers are identified by comparison of measured proteomic values between appropriate case and control populations. We further analyze candidate biomarkers by, for example, assessing their biological plausibility, which can provide additional information and confidence in selecting biomarkers for developing diagnostic algorithms. We also apply analytical tools that we have developed to assess sample quality and to detect potential biases in the data, such as preanalytical variability (unpublished).

8.4.5 Algorithm Development

Once a set of candidate biomarkers is selected, we use them in a process of training and testing diagnostic algorithms. We and others believe that panels of multiple biomarkers are most likely to provide the most sensitive and specific diagnostic information for a given condition. We typically employ a variety of methods to develop diagnostic algorithms, including Bayesian and random forest approaches (Ostroff et al. 2010b). A common problem encountered in algorithm development is so-called overfitting, where the model is tailored too closely to fit the training data and performs worse with new data. To avoid overfitting during algorithm development, we use common strategies including cross-validation during algorithm training and testing the final algorithms with a separate and blinded test set (a predefined subset of samples randomly selected from the study and blinded) (Pepe et al. 2008).

8.4.6 Translating Diagnostic from Discovery to Clinical Format

Moving a diagnostic from the discovery platform to the final clinical format is a critical step that has often failed due to differences in measurement technologies. For example, many candidate biomarkers discovered by MS fail to translate to ELISAs due to differences between the technologies (Rifai et al. 2006). A key advantage of our technology is that we can use the same affinity reagents (SOMAmers) for biomarker discovery and the clinical assay format. For example, we have developed a streamlined version of our SOMAmer-based assay using streptavidin-coated microtiter plates and plate washing equipment that is standard in clinical labs (Kraemer et al. 2011). The streamlined platform is capable of efficiently measuring panels of a few to a few hundred proteins with performance similar to our discovery platform (Kraemer et al. 2011). In the course of developing diagnostics, we have successfully transferred diagnostic tests from the discovery platform to the streamlined platform for conducting validation studies (Kraemer et al. 2011).

8.4.7 Pivotal Evaluation and Validation

The final phases of developing a diagnostic are analogous to those of drug development and rely on a pivotal evaluation of the diagnostic in a clinical study designed to validate the test in an independent set of samples (Pepe et al. 2008). There are examples of pivotal evaluation of diagnostics using both retrospective and prospective study designs. Regardless, this step demands judicious planning and proper guidance and communication with the appropriate regulatory body (e.g., the FDA in the United States). As with drug development, pivotal evaluations of diagnostics are expensive and time-consuming. The pivotal evaluation is designed to provide results to complete the package that supports final submission for regulatory approval.

8.4.8 Regulatory Submission and Commercialization

Seeking regulatory approval is typically the culmination of the diagnostics development process. In most cases, diagnostic tests must be approved by an appropriate regulatory body, such as the FDA, in order for the test to be marketed and sold. This is a complicated process that is well beyond the scope of this review (see Chap. 12) and requires specific expertise to plan and execute. Moreover, the development of a successful diagnostic does not end with regulatory approval, one could argue it begins here. Commercialization is an equally challenging process with its set of considerations, strategies, and experts. Once again, this topic is beyond the scope of this review (see also Chap. 12). Nevertheless, we do mention this because it is essential to consider commercialization from the very beginning of the diagnostic development process. In the following section, we present two examples of diagnostics that we are developing.

8.5 Example # 1: Lung Cancer Diagnostic

We recently published the first results for a diagnostic product that is in commercial development for the early detection of lung cancer (Ostroff et al. 2010b). This was the first large-scale clinical application of our SOMAscan platform and is notable among clinical proteomic biomarker studies for its breadth, scale, and efficiency. In this study, 833 proteins were quantified for each of 1,326 subjects in a highly automated process that took just eight days. This is the equivalent of over 1,000,000 ELISA measurements. The paper (Ostroff et al. 2010b) reports results from the first five phases of our product development process: product concept, study design, biomarker identification, algorithm development, and blinded verification. After this point, development was transferred to a commercial partner and continues into the next phases. Here we summarize the published results.

Lung cancer is the leading cause of cancer deaths because about 84% of cases are diagnosed at an advanced stage (Jemal et al. 2008; Kassis et al. 2009; Okada et al. 2005). Worldwide statistics show that in 2008, ~1.5 million people were diagnosed and ~1.3 million died (Boyle and Levin 2008), and this survival rate is unchanged since 1960. However, patients diagnosed at an early stage and who undergo surgery experience an 86% overall 5-year survival (Kassis et al. 2009; Okada et al. 2005). New diagnostics are therefore needed to identify early-stage lung cancer.

In this study, we conducted a multicenter case–control study of non-small cell lung cancer (NSCLC) in long-term tobacco-exposed populations. We analyzed archived serum samples collected from 1,326 subjects enrolled in four independent studies. Sera from lung cancer cases were collected from 291 patients within 8 weeks of the first biopsy-proven lung cancer and prior to surgical tumor removal. Sera from control cases were collected from 1,035 asymptomatic study participants with $\geq$10 pack years of cigarette smoking.

We measured 813 proteins (the size of the menu at the time of analysis) in each sample, identified 44 candidate biomarkers, and developed a 12-protein panel that discriminates NSCLC from controls with 91% sensitivity and 84% specificity in cross-validated training and 89% sensitivity and 83% specificity in a separate verification set, with similar performance for early- and late-stage NSCLC (Fig. 8.6).

The sensitivity and specificity of the 12-biomarker panel identified improve upon published protein and gene expression panels (Ostroff et al. 2010b). This study is a significant advance in clinical proteomics in an area of high unmet clinical need, and it interrogated a broader sampling of the proteome with a larger dynamic range than previously published clinical studies that broadly profiled the serum proteome with platforms including mass spectrometry, antibody arrays, and

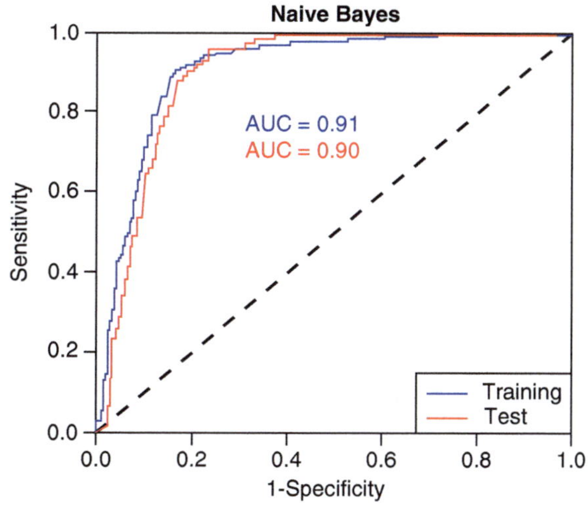

Fig. 8.6 Lung cancer biomarker performance. ROC curve shows performance of a 12-protein biomarker panel for classifying individuals diagnosed with NSCLC from at-risk tobacco-exposed controls. Figure reproduced from Ostroff et al. (2010b)

autoantibody arrays. Based on these discoveries, we are engaged with a commercial partner in developing a clinical blood test to enable an earlier diagnosis of lung cancer. This study was the first published from a sequence of successful biomarker discovery studies we have completed in different areas of unmet medical need.

8.6 Example #2: Chronic Kidney Disease

Another example of a diagnostic product in development is chronic kidney disease (CKD), the slow loss of kidney function over time. We recently published a clinical biomarker study of CKD that identified a number of CKD biomarkers that could help in developing better diagnostics for early detection of CKD, which can lead to early intervention and substantially improved prognosis (Chaudhary et al. 2010; Giannelli et al. 2007; Levey et al. 2007; Nickolas et al. 2008). CKD is a growing global public health epidemic that is "common, harmful, and treatable" with an estimated prevalence of nearly 10% worldwide (Levey et al. 2007). Early diagnosis of CKD requires predictive, noninvasive biomarkers, which could also be useful for monitoring disease progression and guiding treatment.

The kidney filters molecules from serum based on size (molecular mass) and charge (Venturoli and Rippe 2005), and we hypothesized that disease progression would increase the concentration of small plasma proteins (MW <45 kDa). The current clinical practice for diagnosis of CKD is to calculate estimated glomerular filtration rate (eGFR) from measuring the clearance of creatinine, an imperfect biomarker of CKD progression. Creatinine clearance is notably inaccurate in diagnosing early CKD. Therefore, we searched for potential CKD biomarkers by analyzing plasma samples from 42 subjects diagnosed with early- and late-stage CKD. Eleven subjects were diagnosed with early-stage CKD (stages 1–2, median

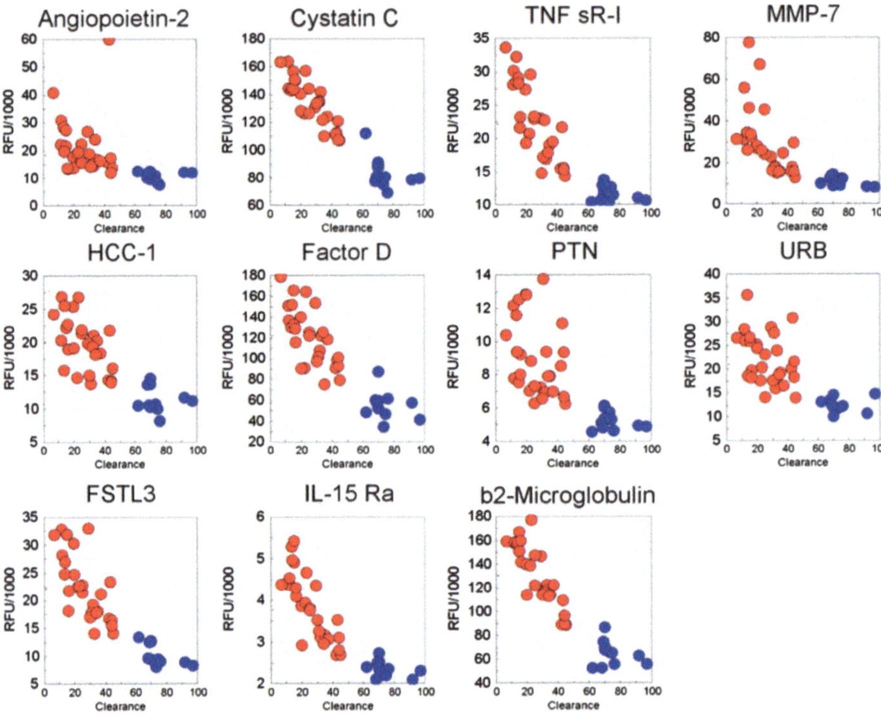

Fig. 8.7 Potential CKD biomarkers. Eleven analytes with the smallest *q*-values ($<3.5 \times 10^{-7}$). Protein concentrations (expressed as RFU values) as a function of renal clearance for the 11 best biomarkers of late-stage (*red circles*) vs. early-stage CKD (*blue circles*). Figure reproduced from Gold et al. (2010a)

creatinine clearance 70 ml/min/m^2, range 62–97 ml/min/m^2), and 31 were diagnosed with late-stage CKD (stages 3–5, median creatinine clearance 25 ml/min/m^2, range 7–49 ml/min/m^2).

We measured 614 human proteins (the assay menu at the time the analysis was conducted) in multiplex for each sample. We compared the results from early- to late-stage CKD subjects and identified 60 proteins that varied significantly between the two groups [Mann–Whitney test, *p*-value $<4.2 \times 10^{-4}$ after correcting for multiple tests (FDR cutoff of $<4.2 \times 10^{-4}$)]. Eleven proteins with the most significant variation between early- and late-stage CKD are highlighted in Fig. 8.7. As hypothesized, the majority of these proteins (all except angiopoietin-2 and URB) are relatively small (<25 kDa). Estimated GFR was inversely correlated with protein concentration, which suggests they could be potential biomarkers of CKD progression. Among the potential CKD biomarkers identified, cystatin C and β_2-microglobulin are important and previously known (Chaudhary et al. 2010; Giannelli et al. 2007; Nickolas et al. 2008). Furthermore, the proteins complement factor D and TNF sR-I have also been found elevated in CKD (Pascual et al. 1988; van Riemsdijk-van Overbeeke et al. 2000). Overall, the results of this study showed many strong potential biomarkers for CKD and good promise for developing new diagnostics to help diagnose CKD early and improve prognosis of this serious and growing epidemic.

8.7 Conclusion

Microarrays have played a critical role in the evolution and ultimate success of our proteomics technology. Together, these published and unpublished results demonstrate the power of the SOMAscan proteomic technology for discovering robust biomarkers for developing new products for personalized medicine. Our general approach of unbiased biomarker discovery can be applied to many more conditions including infectious, inherited, neurological, and metabolic diseases. The SOMAscan platform is enabling a broad pipeline of diagnostics across many diseases.

Acknowledgments We thank all our colleagues who have contributed to developing SomaLogic's unbiased, high-scale proteomics technology and in particular those who developed SOMAmers and the SOMAscan proteomics assay for biomarker discovery. We especially thank Nebojsa Janjic, Nick Saccomano, and Steve Williams and their research groups for their dedicated efforts in developing this technology.

References

Addona TA, Abbatiello SE, Schilling B et al (2009) Multi-site assessment of the precision and reproducibility of multiple reaction monitoring-based measurements of proteins in plasma. Nat Biotechnol 27:633–641. doi:10.1038/nbt.1546

Aebersold R (2009) A stress test for mass spectrometry–based proteomics. Nat Methods 6:411–412. doi:10.1038/nmeth0609-411

Anderson NL (2002) The human plasma proteome: history, character, and diagnostic prospects. Mol Cell Proteomics 1:845–867. doi:10.1074/mcp.R200007-MCP200

Anderson NL (2010) The clinical plasma proteome: a survey of clinical assays for proteins in plasma and serum. Clin Chem 56:177–185. doi:10.1373/clinchem.2009.126706

Bell AW, Deutsch EW, Au CE et al (2009) A HUPO test sample study reveals common problems in mass spectrometry-based proteomics. Nat Methods 6:423–430. doi:10.1038/nmeth.1333

Bock C, Coleman M, Collins B et al (2004) Photoaptamer arrays applied to multiplexed proteomic analysis. Proteomics 4:609–618. doi:10.1002/pmic.200300631

Boyle P, Levin B (2008) World cancer report. International Agency for Research on Cancer, Lyon

Brody EN, Gold L, Lawn RM et al (2010) High-content affinity-based proteomics: unlocking protein biomarker discovery. Expert Rev Mol Diagn 10:1013–1022. doi:10.1586/erm.10.89

Chaudhary K, Phadke G, Nistala R et al (2010) The emerging role of biomarkers in diabetic and hypertensive chronic kidney disease. Curr Diab Rep 10:37–42. doi:10.1007/s11892-009-0080-z

Eaton BE, Gold L, Zichi DA (1995) Let's get specific: the relationship between specificity and affinity. Chem Biol 2:633–638. doi:10.1016/S0031-9406(05)67026-5

Ellington AD, Szostak JW (1990) In vitro selection of RNA molecules that bind specific ligands. Nature 346:818–822. doi:10.1038/346818a0

Giannelli SV, Patel KV, Windham BG et al (2007) Magnitude of underascertainment of impaired kidney function in older adults with normal serum creatinine. J Am Geriatr Soc 55:816–823. doi:10.1111/j.1532-5415.2007.01196.x

Glass DJ (2010) A critique of the hypothesis, and a defense of the question, as a framework for experimentation. Clin Chem 56:1080–1085. doi:10.1373/clinchem.2010.144477

Gold L, Ayers D, Bertino J et al (2010a) Aptamer-based multiplexed proteomic technology for biomarker discovery. PLoS One 5:e15004. doi:10.1371/journal.pone.0015004

Gold L, Janjic N, Jarvis T et al (2010b) Aptamers and the RNA world, past and present. In: Atkins JF, Gesteland RF, Cech T (eds) RNA worlds, 3rd edn. Cold Spring Harbor, NY, pp 343–341

Golden MC, Collins BD, Willis MC, Koch TH (2000) Diagnostic potential of PhotoSELEX-evolved ssDNA aptamers. J Biotechnol 81:167–178

Hopfield JJ (1974) Kinetic proofreading: a new mechanism for reducing errors in biosynthetic processes requiring high specificity. Proc Natl Acad Sci USA 71:4135–4139

Hopfield JJ, Yamane T, Yue V, Coutts SM (1976) Direct experimental evidence for kinetic proofreading in amino acylation of tRNAIle. Proc Natl Acad Sci USA 73:1164–1168

Huang D-B, Vu D, Cassiday LA et al (2003) Crystal structure of NF-kappaB (p50)2 complexed to a high-affinity RNA aptamer. Proc Natl Acad Sci USA 100:9268–9273. doi:10.1073/pnas.1632011100

Jemal A, Siegel R, Ward E et al (2008) Cancer statistics, 2008. CA Cancer J Clin 58:71–96. doi:10.3322/CA.2007.0010

Kassis ES, Vaporciyan AA, Swisher SG et al (2009) Application of the revised lung cancer staging system (IASLC Staging Project) to a cancer center population. J Thorac Cardiovasc Surg 138:412–418. doi:10.1016/j.jtcvs.2009.01.033, e1-2

Keeney TR, Bock C, Gold L et al (2009) Automation of the SomaLogic proteomics assay: a platform for biomarker discovery. J Assoc Lab Automation 14:360–366. doi:10.1016/j.jala.2009.05.003

Kell DB, Oliver SG (2004) Here is the evidence, now what is the hypothesis? the complementary roles of inductive and hypothesis-driven science in the post-genomic era. Bioessays 26:99–105. doi:10.1002/bies.10385

Kraemer S, Vaught JD, Bock C et al (2011) From SOMAmer-based biomarker discovery to diagnostic and clinical applications: a SOMAmer-based, streamlined multiplex proteomic assay. PLoS One 6:e26332. doi:10.1371/journal.pone.0026332

Lander AD (2010) The edges of understanding. BMC Biology 8:40. doi:10.1186/1741-7007-8-40

Levey AS, Atkins R, Coresh J et al (2007) Chronic kidney disease as a global public health problem: approaches and initiatives - a position statement from kidney disease improving global outcomes. Kidney Int 72:247–259. doi:10.1038/sj.ki.5002343

Lin H, Lee E, Hestir K et al (2008) Discovery of a cytokine and its receptor by functional screening of the extracellular proteome. Science(80-) 320:807–11. doi:10.1126/science.1154370

Liotta LA, Petricoin EF (2010) Mass spectrometry-based protein biomarker discovery and measurement: sensitivity is the greatest hurdle. Clinical Proteomics 6:4–5. doi:10.1007/s12014-010-9045-0

Long SB, Long MB, White RR, Sullenger BA (2008) Crystal structure of an RNA aptamer bound to thrombin. RNA 14:2504–2512. doi:10.1261/rna.1239308

Mitchell P (2010) Proteomics retrenches. Nat Biotechnol 28:665–670. doi:10.1038/nbt0710-665

Nabel GJ (2009) The Coordinates of truth. Science 326:53–54

Nickolas TL, Barasch J, Devarajan P (2008) Biomarkers in acute and chronic kidney disease. Curr Opin Nephrol Hypertens 17:127–132. doi:10.1097/MNH.0b013e3282f4e525

O'Farrell PH (1975) High resolution two-dimensional electrophoresis of proteins. J Biol Chem 250:4007–4021

Okada M, Nishio W, Sakamoto T et al (2005) Effect of tumor size on prognosis in patients with non-small cell lung cancer: the role of segmentectomy as a type of lesser resection. J Thorac Cardiovasc Surg 129:87–93. doi:10.1016/j.jtcvs.2004.04.030

Ostroff R, Foreman T, Keeney TR et al (2010a) The stability of the circulating human proteome to variations in sample collection and handling procedures measured with an aptamer-based proteomics array. J Proteomics 73:649–666. doi:10.1016/j.jprot.2009.09.004

Ostroff RM, Bigbee WL, Franklin W et al (2010b) Unlocking biomarker discovery: large scale application of aptamer proteomic technology for early detection of lung cancer. PLoS One 5: e15003. doi:10.1371/journal.pone.0015003

Pan S, Aebersold R, Chen R et al (2009) Mass spectrometry based targeted protein quantification: methods and applications. J Proteome Res 8:787–797. doi:10.1021/pr800538n

Pascual M, Steiger G, Estreicher J et al (1988) Metabolism of complement factor D in renal failure. Kidney Int 34:529–536

Pepe MS, Feng Z (2011) Improving biomarker identification with better designs and reporting. Clin Chem 1095:1093–1095. doi:10.1373/clinchem.2011.164657

Pepe MS, Etzioni R, Feng Z et al (2001) Phases of biomarker development for early detection of cancer OF. Cancer 93:1054–1061

Pepe MS, Feng Z, Janes H et al (2008) Pivotal evaluation of the accuracy of a biomarker used for classification or prediction: standards for study design. J Natl Cancer Inst 100:1432–1438. doi:10.1093/jnci/djn326

Petach H (2002) Dimensionality is the issue: use of photoaptamers in protein microarrays. Curr Opin Biotechnol 13:309–314. doi:10.1016/S0958-1669(02)00329-4

Ransohoff DF (2004) Rules of evidence for cancer molecular-marker discovery and validation. Nat Rev Cancer 4:309–314. doi:10.1038/nrc1322

Ransohoff DF, Gourlay ML (2010) Sources of bias in specimens for research about molecular markers for cancer. J Clin Oncol 28:698–704. doi:10.1200/JCO.2009.25.6065

Rifai N, Gillette MA, Carr SA (2006) Protein biomarker discovery and validation: the long and uncertain path to clinical utility. Nat Biotechnol 24:971–983. doi:10.1038/nbt1235

Service RF (2008) Proteomics ponders prime time. Science (80-) 321:1758–1761

Silberring J, Ciborowski P (2010) Biomarker discovery and clinical proteomics. Trends Anal Chem 29:128. doi:10.1016/j.trac.2009.11.007

Smith D (2002) Sensitivity and specificity of photoaptamer probes. Mol Cell Proteomics 2:11–18. doi:10.1074/mcp.M200059-MCP200

Smith D, Collins BD, Heil J, Koch TH (2003) Sensitivity and specificity of photoaptamer probes. Mol Cell Proteomics 2:11–18

Tuerk C, Gold L (1990) Systematic evolution of ligands by exponential enrichment: RNA ligands to bacteriophage T4 DNA polymerase. Science 249:505–510

van Riemsdijk-van Overbeeke IC, Baan CC, Hesse CJ et al (2000) TNF-alpha: mRNA, plasma protein levels and soluble receptors in patients on chronic hemodialysis, on CAPD and with end-stage renal failure. Clin Nephrol 53:115–123

Vaught JD, Dewey T, Eaton BE (2004) T7 RNA polymerase transcription with 5-position modified UTP derivatives. J Am Chem Soc 126:11231–11237. doi:10.1021/ja049009h

Vaught JD, Bock C, Carter J et al (2010) Expanding the chemistry of DNA for in vitro selection. J Am Chem Soc 132:4141–4151. doi:10.1021/ja908035g

Venturoli D, Rippe B (2005) Ficoll and dextran vs. globular proteins as probes for testing glomerular permselectivity: effects of molecular size, shape, charge, and deformability. Am J Physiol Renal Physiol 288:F605–F613. doi:10.1152/ajprenal.00171.2004

Vicens Q, Cech TR (2006) Atomic level architecture of group I introns revealed. Trends Biochem Sci 31:41–51. doi:10.1016/j.tibs.2005.11.008

Zichi D, Koga T, Greef C et al (2002) Photoaptamer technology: development of multiplexed microarray protein assays. Clin Chem 48:1865–1867

Zichi D, Eaton B, Singer B, Gold L (2008) Proteomics and diagnostics: let's get specific, again. Curr Opin Chem Biol 12:78–85. doi:10.1016/j.cbpa.2008.01.016

Part III
Other Types of Arrays

Chapter 9
Tissue Microarrays for Translational Research

Ronald Simon and Martina Mirlacher

9.1 Introduction

The decryption of the human genome in 2001 and the broad availability of DNA microarrays during the last decade have opened a new era in biomedical research. Since then, thousands of studies have employed array-based high-density RNA expression or DNA copy number profiling analysis in order to identify genes whose structure or expression is altered in pathological tissues. Currently, large international research consortia like the International Cancer Genome Consortium (http://www.ICGC.org) aim at whole genome sequencing of normal and pathological human tissues. As a consequence, a huge number of genes have been (and still will be) identified that may include promising target genes for new therapies. In order to translate basic genetic findings into clinically useful applications, a thorough validation of the identified alterations in clinically well-defined tissues is a crucial step. Collecting data on molecular epidemiology is often a significant challenge for basic researchers due to limited availability of human tissues with clinical information. The tissue microarray (TMA) technology greatly facilitates large-scale tissue analyses. With this method, up to 1,000 different minute tissue samples can be brought on one microscope glass slide and then be simultaneously analyzed by all types of in situ methods. Although its roots go back to the 1980s when researchers first started to combine several small pieces of tissues into so-called sausage blocks (Battifora 1986), today's TMA technique has a number of advantages. Firstly, the cylindrical shape and the small diameter of the specimen that is taken out of the donor block minimize the damage to the tissue and avoid depleting the original tissue blocks. This allows all kinds of morphological and molecular analyses that may subsequently become necessary on the original tissue

R. Simon (✉) • M. Mirlacher
Institute of Pathology, University Medical Center Hamburg Eppendorf, Martinistrasse 52, Hamburg, Germany
e-mail: r.simon@uke.de

B. Jordan (ed.), *Microarrays in Diagnostics and Biomarker Development*,
DOI 10.1007/978-3-642-28203-4_9,

blocks. Secondly, and probably most importantly, the true novelty of the TMA technology is the linkage of histological and appendant clinical data of the tissues on one slide. The very high number of tissues, the small size and regular shape of the tissue spots, and the highly organized array pattern made it a discrete technique with unique features.

Since its first publication in 1998 (Kononen et al. 1998), the TMA technology has imposingly demonstrated that it is more than just an efficient and economic alternative to expensive and time-consuming analysis of conventional large section tissue sections. Today, tissue microarrays have become a standard tool allowing for a new dimension of tissue analysis with a previously inaccessible statistical power.

9.2 Methods for TMA Construction

Most of the work related to TMA manufacturing is traditional pathology work similar to what is needed for conventional studies involving "large" tissue sections. The availability of a large collection of well-characterized tissues—optimally with attached clinical data—is the most important prerequisite. About 95% of the work related to the manufacturing process is the collection of potentially relevant tissues, reviewing the corresponding slides, and selecting blocks for subsequent arraying. The time needed for the block collection greatly depends on the degree of organization of a tissue archive and its related databases.

Many different techniques have been developed to construct tissue microarrays. The *recipient block technology* (Kononen et al. 1998; Mirlacher and Simon 2010) is the "classical" and by far most widely used method of TMA making (Fig. 9.1). With this method, minute cylindrical tissue punches typically measuring 0.6 mm in diameter are removed from donor tissue blocks and are transferred into empty "recipient" paraffin blocks with precored or predrilled holes to hold the punches. The equipment needed to make TMAs is not complicated. The key components of commercially available tissue microarraying devices are two needles with a slightly different diameter combined with a mechanism allowing the introduction of holes at defined areas (coordinates) in a recipient block. With the smaller needle (outer diameter 0.6 mm, this needle may also be replaced by a 0.6 mm drill), holes are punched (or drilled) into empty "recipient" paraffin blocks. Subsequently, a slightly larger needle (inner diameter 0.6 mm) is utilized to transfer tissue cylinders from preexisting "donor" paraffin blocks into these premade holes at specific coordinates. Up to 1,000 different tissues can be analyzed in one 2 cm by 3 cm TMA block. Although there is no general rule to the "optimal" size of the tissue spots, a diameter of 0.6 mm has become a standard size for tumor analysis. Larger core diameters (for instance 1.0–1.2 mm) are advisable for analysis of normal tissues, in order to make sure that all relevant cell types of the organ of interest are represented in a single tissue core.

The *cutting-edge matrix assembly* (CEMA) arraying method combines serial cutting and edge-to-edge bonding of samples to assemble a scaffold-free array

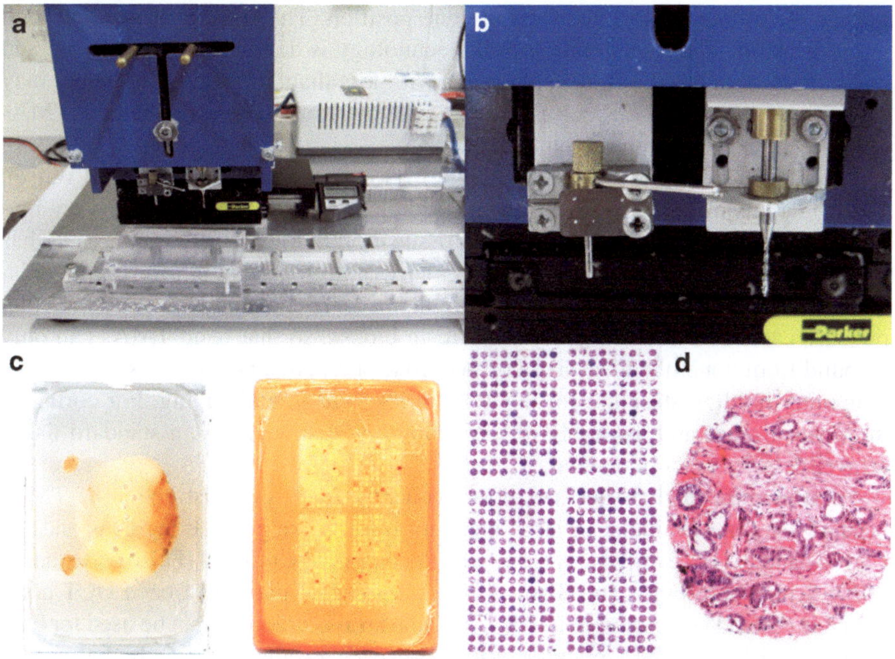

Fig. 9.1 Tissue microarrayer. (**a**) Custom-made arrayer device including a hollow needle to transfer tissue cores from the donor block and a drill to prepare holes in the recipient block. (**b**) Close-up showing the hollow needle and the drill. (**c**) Donor block with holes where the tissue cores have been removed, and the completed TMA block containing 540 tissue spots. (**d**) Overview of a hematoxylin and eosin stained array section and magnification of a 0.6-mm breast cancer tissue spot

matrix (LeBaron et al. 2005; Rui and LeBaron 2005). Building of the array blocks involves a two-step procedure of repeated cutting and bonding of ultrathick sections (e.g., 100 μm). The section thickness determines the size of the rectangular array "spot" present in the final CEMA array. Since no paraffin matrix is necessary to build the array, and because very small spot dimensions can be implemented, up to 10,000 tissues can be placed on one microscopic slide. Probably the most important benefit of the CEMA technology is that no special equipment (except a regular microtome) is required, making this technology an attractive alternative to the standard recipient array technology in laboratories with low budget.

Likewise, the *hypodermic needle technology* does not require specific equipment (Pires et al. 2006; Pires and de Souza 2010). Hollow needles of standard diameters are made from customized commercial 16–21-gauge needles with inner diameters between 0.5 and 1.2 mm, where the original bevel is removed and the inner surface is sharpened. Tissue cores are taken from the donor blocks and removed from the needle with forceps. The punches are then manually attached to a double-sided tape that has been placed in a standard (3 cm by 4 cm) stainless steel mold used for paraffin block making. This procedure requires special diligence and a perfectly flat bottom of the punches to avoid an oblique position. The mold is then filled with liquid paraffin. The hypodermic needle technology is definitely an affordable and quick method to build low-density arrays with less than 100 tissue cores, but may become comparatively difficult if aiming in the construction of high-density TMAs with little space between the punches.

In the *resin tissue microarray technology* (Howat et al. 2005; Howat and Wilson 2010), the advantage of resin as an embedding medium that allows for very thin (down to 0.5 μm) sections with excellent tissue morphology (Britten et al. 1993) has been combined with the improved antigenicity of acetone fixation for immunohistochemistry analysis. An important limitation of the resin technology is that punches cannot be cored from existing resin blocks, so that resin TMAs can only be build from nonembedded (fixed or frozen) tissues. An agarose block is used as a temporary holding structure for the TMA prior to resin embedding. For coring of the donor tissues and assembly of the temporary agarose array, a standard tissue arrayer device is recommended.

Paraffin-embedded tissues are most frequently used for construction of TMAs because they are available in large numbers from existing tissue archives and are easy to handle. However, *frozen tissue samples* can also be utilized (Fejzo and Slamon 2010; Schoenberg Fejzo and Slamon 2001). Instead of paraffin, frozen OCT compound is utilized as a carrier material. The same arraying devices can be used for both paraffin and frozen TMA manufacturing. However, needles are at a higher risk of bending and breaking than in case of paraffin-embedded tissues. Frozen arrays are highly advantageous if paraffin suitable antibodies are not available for the target protein, and are favored for RNA in situ expression analysis because of better RNA preservation. Frozen TMAs may also be superior for fluorescence-based protein quantification experiments because of lower autofluorescence as compared to formalin-fixed tissues.

TMAs may also be made from *core needle biopsies*, allowing for efficient analysis of precious tissues that are only available in limited amounts. For example, in prostate cancer, biopsies often represent the sole material for histological diagnosis, since many patients are not treated by radical prostatectomy. Such biopsies typically measure only 0.6 mm in diameter and 5–20 mm in length, with a total volume of less than 20 mm^3. Several researchers have successfully manufactured TMAs from core needle biopsies (Datta et al. 2005; Datta and Kajdacsy-Balla 2010; Jhavar et al. 2005; McCarthy et al. 2009).

Finally, TMAs can also be made *from experimental tissues and cells in suspension* like cell lines (Montgomery et al. 2005; Simon et al. 2002) (Fig. 9.2) or xenografts (Bubendorf et al. 1999). Constructing TMAs from suspension cells is particularly useful for analysis of hematopoietic neoplasms, fine needle aspirates, and body fluids (Zhao and Natkunam 2010). Cell line TMAs containing large numbers of different cell lines are useful for selection of cell lines with distinct genetic features. Suitable cell lines can then be ordered, cultured, and, for example, used for functional arrays or serve as model systems for the analysis of particular molecular pathways.

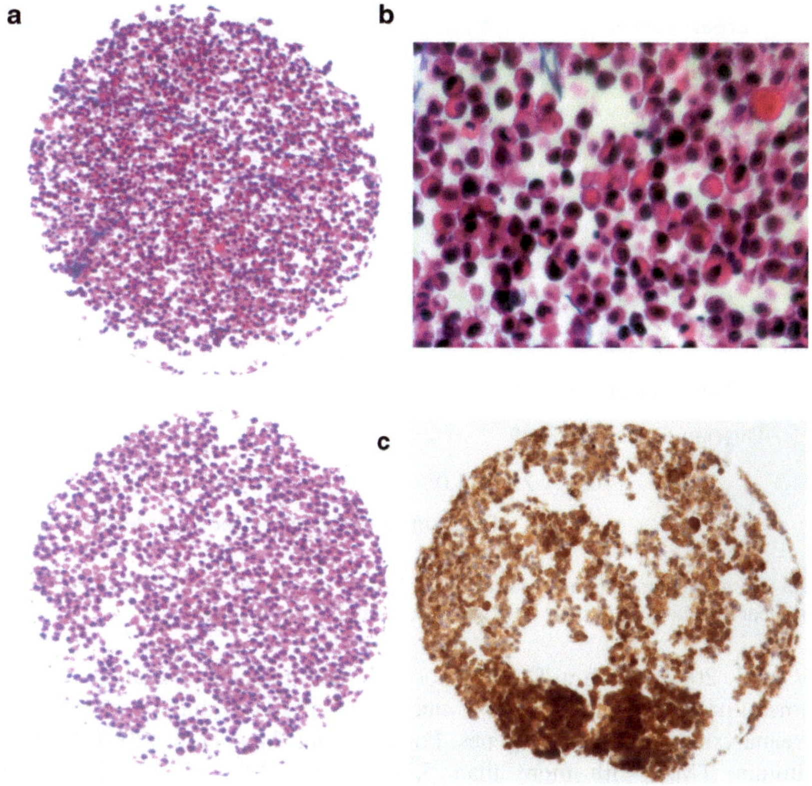

Fig. 9.2 Cell line TMA. (**a**) Overview of two spots made from breast cancer cell lines. (**b**) Magnification showing hematoxylin and eosin stained cells. (**c**) Immunohistochemistry of a breast cancer cell line spot showing strong cytoplasmic staining

9.3 Tissue Microarray Applications

The vast majority of published studies have used TMAs in cancer research, but there are a growing number of studies employing the TMA technology also for research in nonneoplastic tissues, for example, in inflammatory and neurodegenerative disease. Kellner et al. (2009) constructed a large Alzheimer disease (AD) TMA containing 2,325 tissue specimens from three defined regions of 48 AD patients and 48 age-matched controls. In cancer research, different types of TMAs can be manufactured with an optimized composition of the tissue set and on the dataset attached to the tissues, including prevalence TMAs, progression TMAs, prognostic TMAs, heterogeneity TMAs, and TMAs from normal tissues.

Prevalence TMAs contain tumor samples from one or several tumor entities without extensive clinicopathological information. Prevalence TMAs containing tissue samples from various different tumor entities can be used to comprehensively study the epidemiology of molecular features in human neoplasia (Schraml et al. 1999;

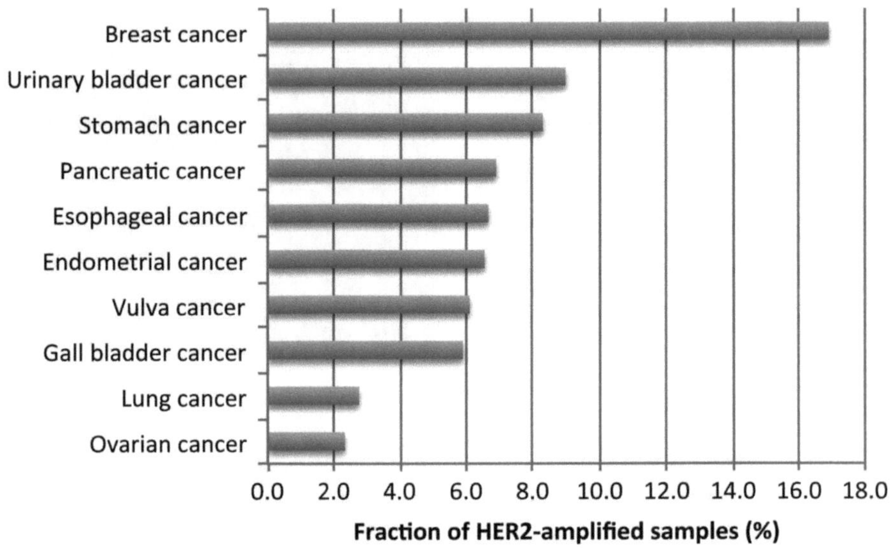

Fig. 9.3 Results of a multitumor array analysis showing the prevalence of HER2 amplification in ten different human cancer types. Data were taken from Tapia et al. (2007)

Went et al. 2004). Such a TMA is not only instrumental to analyze expression patterns of newly detected genes but can also be of great value for the comprehensive reanalysis of well-known genes. For example, Tapia et al. (2007) analyzed a multitumor TMA with more than 5,700 samples for HER2, the target of trastuzumab (Herceptin), a monoclonal antibody that has been approved for treatment of HER2-amplified or overexpressing breast cancer in 1998. In this study, gene amplification was assessed by fluorescence in situ hybridization and Her2 overexpression by immunohistochemistry. Although Her2 overexpression (score 2/3+) and amplification occurred most often in breast cancers, it was also found in 18 other tumor types including cancers of the urinary bladder, stomach, endometrium, lung, and ovaries (Fig. 9.3). Such data can be most valuable to guide research and pharmaceutical product development. For example, in 2010, Herceptin was also approved for treatment of HER2-positive stomach cancer. Smaller multitumor arrays have been used to study the expression of potential new tumor markers such as SPANX-B (Almanzar et al. 2009), Sil (Erez et al. 2004), NOX1 (Geiszt et al. 2003), COX2, MMP2, or MMP9 (Dicken et al. 2006).

Progression TMAs are constructed from samples of different stages of a particular tumor type. For example, we have recently built a prostate cancer progression TMA composed of normal prostate tissues, prostatic intraepithelial hyperplasia (PIN), untreated localized tumors, local recurrences, nodal metastases, and distant metastases from late-stage hormone-refractory cancers. Another progression TMA was used to demonstrate a high concordance of HER2 amplification and overexpression between primary tumors and their nodal metastases (Simon et al. 2001b).

Prognosis TMAs contain samples from retrospectively collected tumors with long-term clinical follow-up data. They are instrumental to estimate the impact of a given molecular marker on the course of the disease, for example, on overall survival or time to progression. Many studies have utilized prognosis TMAs to reproduce known associations or to find new links between molecular findings and clinical outcome. For example, early studies reported significant associations between estrogen, progesterone expression (Torhorst et al. 2001), or HER-2 alterations (Barlund et al. 2000) and survival in breast cancer patients, or between Ki67 labeling index (LI) and prognosis in urinary bladder cancer (Nocito et al. 2001). In recent years, we used prognosis TMAs to test the clinical impact of oncogene amplification in breast and urinary bladder cancer [including HER2 and TOP2A (Simon et al. 2003), EGFR, CCND1, MYC (Al-Kuraya et al. 2004), FGFR1 and RAF1 (Simon et al. 2001c), MDM2, GLI, CDK4 (Simon et al. 2002), and E2F3 (Oeggerli et al. 2004)], or to validate expression of candidate prognostic markers in prostate cancer (p53 (Schlomm et al. 2008), MYC (El Gammal et al. 2010), EGFR (Schlomm et al. 2007), HER2 (Minner et al. 2010), CD117 (Fleischmann et al. 2009), or CD10 (Fleischmann et al. 2008), (Fig. 9.4). Results from prognosis TMAs are particularly valuable if the best target(s) for drug development must be selected from a series of candidate genes. Prognosis TMAs may also be constructed from tumor samples of patients who were included in clinical trials with standardized treatment. For example, 2,600 breast cancer samples that were treated with trastuzumab and analyzed for HER2 amplification and overexpression by the Breast Cancer International Research Group 006 (BCIRG) have been archived in a TMA format (Press et al. 2005). This TMA will be very effective to study the impact of new marker proteins on the response to treatment with trastuzumab.

Heterogeneity TMAs are made from different areas of the same tumors. They are designed to study intratumoral heterogeneity of molecular markers. For example, (Minner et al. 2012, manuscript in preparation) have constructed a prostate cancer heterogeneity TMA including tumor tissue punches from ten distant areas each of 190 prostatectomy specimens (Fig. 9.5). This TMA was analyzed for expression of the TMPRSS2/ERG fusion gene by means of immunohistochemistry. The results of the study demonstrated that more than 40% of ERG-positive prostate cancers show intratumoral heterogeneity of this marker. Based on the fraction of tissue spots per tumor that show a positive result, heterogeneity TMAs are also suited to estimate the sequence of molecular events during tumor progression. A positive finding in all samples of one tumor argues for an early event, whereas incidental positivity highlights late events. Since targeted therapies will be particularly efficient for early alterations that are present in the majority of tumor cells, heterogeneity TMAs will be very helpful to predict the success of new gene-specific drugs.

Normal tissue arrays are made from healthy tissues. Constructing normal tissue arrays is a particular challenge since healthy tissues are not extensively available in a pathology archive. Normal tissue analysis allows for a comprehensive knowledge on the expression status of a given gene not only on the organ level but also at the level of single cell types. This knowledge is not only important for a better

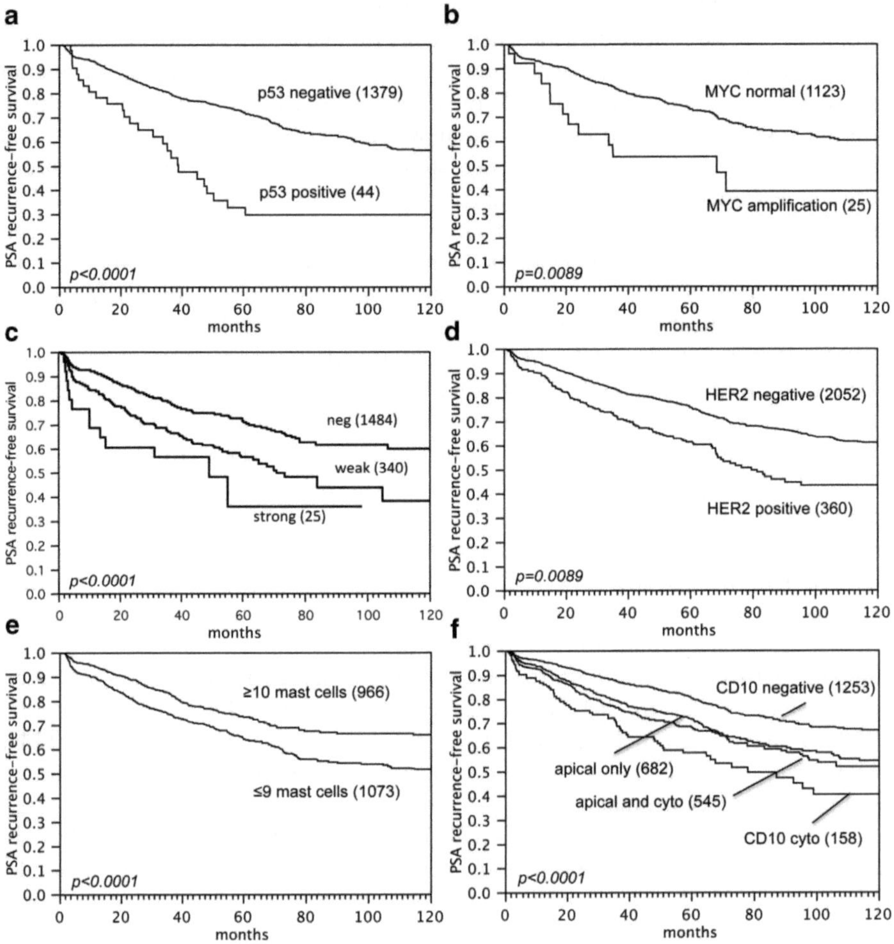

Fig. 9.4 Prognostic relevance of different molecular markers in prostate cancer. Analysis of a 3,261 spots prognosis TMA. (**a**) Kaplan–Meier survival plot showing adverse prognosis of 44 cancers with nuclear accumulation of p53 in comparison to 1,379 cancers without immunohistochemically detectable p53 positivity, (**b**) MYC FISH analysis, (**c**) EGFR immunohistochemistry, (**d**) HER2 immunohistochemistry, (**e**) number of mast cells in the tissue spot as determined by CD117 (KIT) immunohistochemistry, (**f**) CD10 expression pattern. Apical = staining of the apical cell pole

understanding of the gene's role in cancer biology but is also a crucial issue in drug development. Strong expression of a potential drug target protein in essential normal tissues, like brain or heart, is a possible "no-go" criterion for drug development. Human normal tissue TMAs are therefore especially helpful to estimate drug safety before entering into clinical phases of drug testing and can be used to facilitate the process of normal tissue reactivity testing.

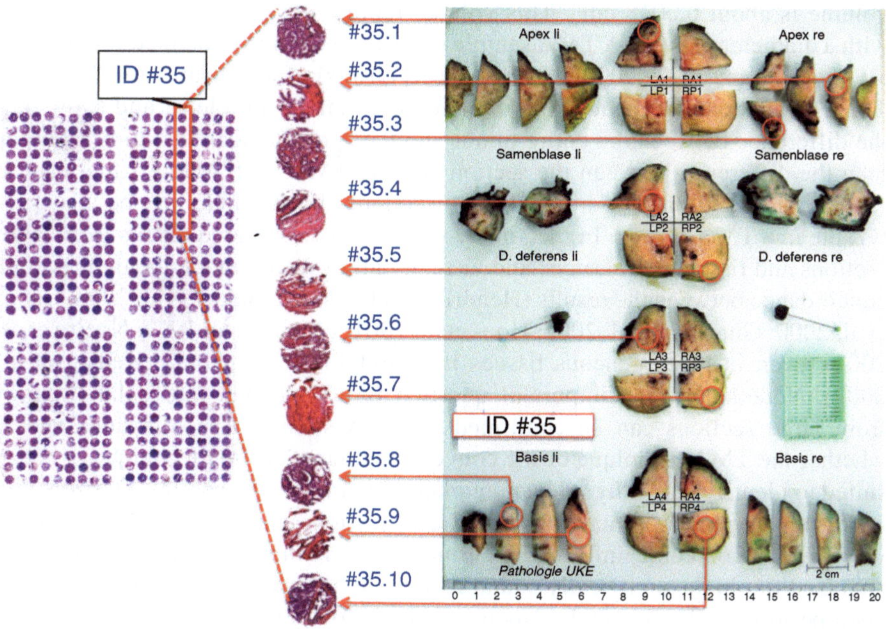

Fig. 9.5 Heterogeneity TMA construction. Whole formalin-fixed cancerous prostates of 190 patients were dissected into 30–50 paraffin blocks. The location of every block was noted before histological examination. Ten tumor tissue cores were taken from ten distant sites of each prostate and used for TMA construction

9.4 Representativity of TMA Studies

When the TMA technology was first published in 1998 (Kononen et al. 1998), the small diameter of the tissue spots (0.6 mm) prompted concerns regarding the representativity of the TMA cores (Camp et al. 2000; Gancberg et al. 2002; Rimm et al. 2001). Tumor heterogeneity is a consequence of genetic instability and molecular evolution, and genetic differences have been found between different cells of the same tumor (Khalique et al. 2007), between invasive and noninvasive parts (Sauter and Mihatsch 1998; Simon et al. 1998; Robanus-Maandag et al. 2003), between primary tumors and metastases (Cho et al. 2008), or between multifocal tumors from the same organ (Simon et al. 2001a). Consequently, many researchers recommended the use of larger core diameters, or the inclusion of more than one tissue spot per tumor to increase the degree of representativity (Camp et al. 2000; Engellau et al. 2001; Fernebro et al. 2002; Hoos et al. 2001a, b, 2002; Kristiansen et al. 2008; Zhang et al. 2003). However, these concerns are based on the rather erroneous assumption that conventional whole sections—the "gold standard" for molecular tumor tissue analysis—are representative of the entire tumor bulk. A typical "large" section contains tumor tissue measuring an area of about 3 × 2 cm. Given a section thickness of 3 μm, the examined tumor volume is about 0.0018 cm^3. This volume represents only 1/19,000 of a tumor with a diameter of 4 cm. A TMA sample measuring 0.6 mm in diameter represents a tumor volume of 0.00000108 cm^3 that is 1/1,600 of a typical "large" section. These figures clearly demonstrate the "representativity problem" and show that the difference between the entire tumor and the "gold standard" section is larger than the difference between the section and the TMA spot. Markers that can be reliably analyzed on conventional large sections should, therefore, be also analyzable in a TMA format. In fact, many studies have compared results from large sections and tissue microarrays, and virtually all of them reported a high level of concordance between the results (Hendriks et al. 2003; Tzankov et al. 2003; Camp et al. 2000; Ginestier et al. 2002; Hoos et al. 2001b; Moch et al. 1999; Nocito et al. 2001), even in heterogeneous tissues like Hodgkin's lymphoma (Hedvat et al. 2002). Nonetheless, the important question is not whether molecular findings from large sections can be confirmed in TMA studies. The relevant issue is whether the TMA technique or the conventional large section technique is better suited to identify clinically relevant markers. In order to answer this question, it is necessary to compare data from both large section analysis and TMA analysis with clinical outcome information. Since large numbers of samples must be analyzed to obtain statistically sound results, only few studies of this kind have been performed. For example, Nocito et al. (2001) demonstrated that histologic grading and Ki67 LI analysis on TMAs and on large sections were equally relevant for predicting prognosis in bladder cancer. Torhorst et al. analyzed the prognostic relevance of p53 expression in a set of 553 breast cancers both on large sections and on four corresponding TMAs made from four different punches of each tumor (Torhorst et al. 2001). Surprisingly, the study yielded discrepant results. Between 15% and 21% of p53-positive tumors were found in the four individual TMAs, and a combination of the results (i.e., a tumor is considered positive if at least one of the four TMAs yielded positive staining) resulted in only 24% p53 positive cases. In contrast, 43% p53-positive tumors were found when the conventional large sections were analyzed, suggesting that a significant fraction of p53-positive tumors had been missed in the TMA analyses. The comparison with follow-up data showed, however, that the TMA results were superior to the large section data, even if only a single TMA spot per donor tissue was analyzed. The expected prognostic difference between p53-positive and p53-negative tumors was found in each of the four TMA analyses, but not in the large section analysis (Fig. 9.6). A reanalysis demonstrated that focal p53 positivity that led to a positive result in 111 tumors according to the large section analysis (but that were all rated negative in the TMA spots) in fact had no biological significance for the clinical course of the tumor.

The above-mentioned study by Torhorst et al. already showed that identical results were obtained for p53 analysis of breast cancers independently of whether one or four tissue spots per tumor were analyzed. In spite of this, the use of four spots per tumor has been suggested when analyzing prostate cancer, since this tumor is often heterogeneous (Rubin et al. 2002). In order to estimate if the

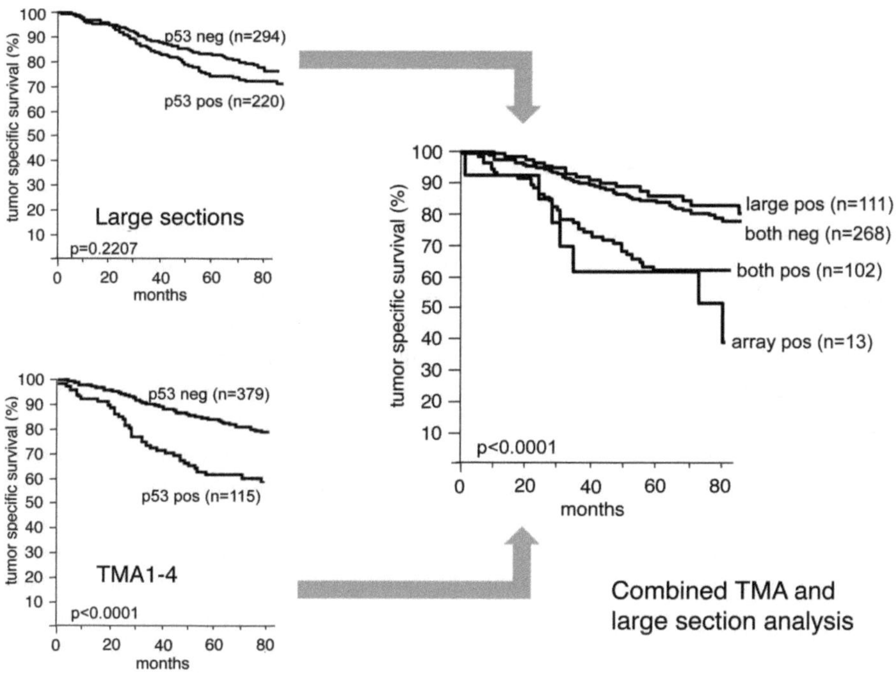

Fig. 9.6 Analysis of p53 immunostaining in >500 breast cancers analyzed on conventional large sections and in a TMA format. The known prognostic relevance was found in the TMA analysis only. A combined reanalysis revealed that 111 cancers that had a positive IHC result in the large section analysis ("large pos") but were negative in the TMA analysis had no negative correlation with patient survival [taken with permission from Sauter (2010)]

additional workload required for analysis and TMA construction with quadruplicate spots is really justified, we recently analyzed Ki67 LI and p53 on three prostate cancer TMAs that were made from cores taken from three different areas of 3,261 prostate cancer tissue blocks (Tennstedt et al. 2012). This study again confirmed that analysis of a single TMA core is sufficient to reproduce the known prognostic significance of p53 and Ki67 LI and that analysis of additional spots did not yield additional prognostic value. The rate of positive findings increased with the amount of analyzed tissue. The average Ki67 LI was higher in tumors with three interpretable TMA spots (LI 5.3) as compared to tumors with only one or two interpretable spots (LI 4.1 each). Similarly, for p53, tumors with three interpretable spots showed positive staining in 3.8% of cases, and tumors with one or two interpretable spots in 1.9% only. These data demonstrate that using multiple TMA cores may also lead to statistical problems since not all tumors will have interpretable tissue on each of the replicate spots, resulting in unequal amounts of tissue analyzed per tumor.

9.5 TMA Analysis

In principle, all analyses applicable to conventional large tissue sections can be done in a TMA format, including immunohistochemistry, fluorescence in situ hybridization (FISH) (Kallioniemi 2010), or RNA in situ (RNA-ISH) analysis (Veeck and Dahl 2010). Most published studies using tissue microarrays (TMAs) have employed immunohistochemistry, which is the "gold standard" method for in situ protein expression ascertainment in tissue samples. Immunohistochemistry analysis of TMA sections can be done with the same protocols used for conventional large section analysis. However, since TMA studies typically include hundreds of tissues and involve a significant workload for the interpreter, every effort should be made to perform the analysis with the highest possible quality. For example, an insufficiently specific antibody will cause high levels of nonspecific background staining that can massively complicate interpretation or even obscure relevant findings. Overdilution of the antibody will result in faint staining and loss of positive findings in case of weak expression of the target protein. To obtain an optimal dynamic range for immunostaining, the primary antibody concentration should be adjusted so that a wide range of staining intensity is observed, optimally with some strongly positive tissue spots, some entirely negative tissue spots, and the remaining tissue spots with variable intermediate staining levels. In addition, immunostaining of formalin-fixed paraffin-embedded (FFPE) tissues often requires antigen retrieval to revert the denaturing effects of the embedding process (Shi et al. 2001). This may involve pretreatment in buffers with different pH, pressure (e.g., incubation in an autoclave), or enzymatic digestion. Since there is no specific and standardized laboratory equipment available for immunohistochemistry, antigen retrieval strategies cannot be easily transferred from one laboratory to another.

A widely used scoring system for immunostaining is to estimate the staining intensity (0, 1+, 2+, 3+) and the percentage of positive tumor cells. From these parameters, a final score is built (see Table 9.1 for an example) that is used for further statistical analysis. Since manual analysis by microscopic inspection of TMA slides is a time-consuming task, several commercial systems have been developed for automated TMA analysis in recent years (Table 9.2). In principle, tissue microarrays are optimally suited for automated analysis (e.g., with automated microscopes or slide scanners) because the most critical step in tissue analysis, i.e., selection of the representative tissue area for scoring, has already been done by an experienced pathologist when the TMA was built. From a purely technical point of view, automated staining quantification can be expected to be more precise than manual reading, but it must be kept in mind that TMA analysis often goes beyond simple staining quantitation. For example, pathologists can easily comprehend the full information provided by a 0.6-mm tissue spot, including subcellular localization of the staining, stained cell types, presence of cancer cells, and also unexpected findings (like staining in blood vessels or lymphocytes), and they will automatically exclude spots lacking tumor cells or showing staining artifacts from the analysis. Automated systems that simply measure staining intensity may not perform

Table 9.1 Scoring of immunohistochemistry

IHC result	Definition
Negative	Int 0
Weak	Int 1 in ≤70% of tumor cells or Int 2 in ≤30% of tumor cells
Moderate	Int 1 in >70% of tumor cells or Int 2 in >30% but ≤70% of tumor cells or Int 3 in ≤30% of tumor cells
Strong	Int 2 in >70% of tumor cells Int 3 in >30% of tumor cells

Int = staining intensity in a 4-step scale (0 = no staining, 1 = low intensity, 2 = intermediate intensity, 3 = high intensity). This scoring may be performed separately for different staining patterns according to the intracellular localization of the staining (membranous, cytoplasmic, nuclear)

Table 9.2 Commercially available automated analysis systems

Company	Product	Web site
Aperio	ScanScope®	http://www.aperio.com
Alphelys	Spot Browser	http://www.alphelys.com
BioGenex	iVision™ and GenoMx™	http://www.biogenex.com
Bioimagene	PATHIAM™	http://www.bioimagene.com
Compucyte Corporation	LSC® and iCyte®	http://www.compucyte.com
Dako Cytomation	ACIS®	http://www.dako.com
Dmetrix	Dx-40	http://www.dmetrix.net
Genetix	Applied Imaging Ariol®	http://www.genetix.com
Hamamatsu	Nanozoomer	http://www.hamamatsu.com
HistoRx	AQUA®/PM2000™	http://www.historx.com
Molecular Devices	Discovery-1™	http://www.moleculardevices.com
Olympus	BLISS HD™	http://olympusamerica.com
SlidePath	OpTMA	http://slidepath.com
TissueGnostics	TissueFAXS	http://www.tissuegnostics.com

optimally in case of proteins with complex functions that may change with their subcellular localization, or that are expressed in more than one cell type. More complex automated analysis systems are capable of identifying regions of interest within tissue spots. For example, the AQUA (automated quantitative analysis) system, which was originally developed at Yale University (McCabe et al. 2005), uses immunofluorescence to colocalize and measure multiple antibodies.

9.6 TMA Sources

A number of commercial and academic vendors offer ready-to-use or custom tissue microarrays (Table 9.3). Most companies offer a wide range of relatively small TMAs comprising of 30–100 donors from various cancerous and normal tissues.

Table 9.3 Selected commercial and academic sources for tissue microarray

Source	Type	Web site
Biomax	Commercial	http://www. biomax.us
BioChain	Commercial	http://www.biochain.com
Foliobest	Commercial	http://www.foliobio.com
HistoBest	Commercial	http://www.histobest.com/
Imgenex	Commercial	http://www.imgenex.com
Cybrdi	Commercial	http://www.cybrdi.com
Petagen	Commercial	http://tissuearray.petagen.com
AMS Biotechnology	Commercial	http://www.amsbio.com
Lifespan Biosciences	Commercial	http://www.lsbio.com/
Protein Biotechnologies	Commercial	http://www.proteinbiotechnologies.com
TriStar Technology Group	Commercial	http://www.tristargroup.us/
Origene	Commercial	http://www.origene.com
National Cancer Institute	Academic	http://cdp.nci.nih.gov/
Yale school of medicine	Academic	http://medicine.yale.edu/pathology/researchprograms/tissueservices/tma.aspx

Some companies offer also arrays constructed from frozen tissue samples, including BioChain®, AMS Biotechnology Ltd., or TriStar Technology Group. Larger TMAs with pathological information and clinical follow-up data can be obtained from academic institutes. For example, the Cancer Diagnosis Program of the National Cancer Institute offers breast cancer TMAs with up to 1,000 donors and colon cancer TMAs from 350 donors.

9.7 Conclusions

Tissue microarrays have opened a new era for tissue analysis. Basically all kinds of in situ analyses, like IHC and in situ hybridization assays, may be adapted to TMAs with only slight (if any) modifications of the corresponding large section protocols. TMAs may be composed in many different ways, emphasizing, for example, prevalence of molecular alterations, or prognostic value in diseased tissues. They have significantly accelerated translational research and play an important role for drug development. In addition, they also offer an attractive possibility for biobanking of tissue samples, for example, from clinical studies. Although an automated analysis is principally feasible, the vast majority of TMAs are still analyzed manually by pathologists, highlighting the roots of the technology, that is an ultimate miniaturization of classical pathology procedures.

References

Al-Kuraya K, Schraml P, Torhorst J, Tapia C, Zaharieva B, Novotny H et al (2004) Prognostic relevance of gene amplifications and coamplifications in breast cancer. Cancer Res 64(23):8534–8540

Almanzar G, Olkhanud PB, Bodogai M, Dell'agnola C, Baatar D, Hewitt SM et al (2009) Sperm-derived SPANX-B is a clinically relevant tumor antigen that is expressed in human tumors and readily recognized by human CD4+ and CD8+ T cells. Clin Cancer Res 15(6):1954–1963

Barlund M, Forozan F, Kononen J, Bubendorf L, Chen Y, Bittner ML et al (2000) Detecting activation of ribosomal protein S6 kinase by complementary DNA and tissue microarray analysis. J Natl Cancer Inst 92(15):1252–1259

Battifora H (1986) The multitumor (sausage) tissue block: novel method for immunohistochemical antibody testing. Lab Invest 55(2):244–248

Britten KM, Howarth PH, Roche WR (1993) Immunohistochemistry on resin sections: a comparison of resin embedding techniques for small mucosal biopsies. Biotech Histochem 68(5):271–280

Bubendorf L, Kolmer M, Kononen J, Koivisto P, Mousses S, Chen Y et al (1999) Hormone therapy failure in human prostate cancer: analysis by complementary DNA and tissue microarrays. J Natl Cancer Inst 91(20):1758–1764

Camp RL, Charette LA, Rimm DL (2000) Validation of tissue microarray technology in breast carcinoma. Lab Invest 80(12):1943–1949

Cho EY, Han JJ, Choi YL, Kim KM, Oh YL (2008) Comparison of Her-2, EGFR and cyclin D1 in primary breast cancer and paired metastatic lymph nodes: an immunohistochemical and chromogenic in situ hybridization study. J Korean Med Sci 23(6):1053–1061

Datta MW, Kajdacsy-Balla AA (2010) Tissue microarrays from biopsy specimens. Methods Mol Biol 664:103–111

Datta MW, Kahler A, Macias V, Brodzeller T, Kajdacsy-Balla A (2005) A simple inexpensive method for the production of tissue microarrays from needle biopsy specimens: examples with prostate cancer. Appl Immunohistochem Mol Morphol 13(1):96–103

Dicken BJ, Graham K, Hamilton SM, Andrews S, Lai R, Listgarten J et al (2006) Lymphovascular invasion is associated with poor survival in gastric cancer: an application of gene-expression and tissue array techniques. Ann Surg 243(1):64–73

El Gammal AT, Bruchmann M, Zustin J, Isbarn H, Hellwinkel OJ, Kollermann J et al (2010) Chromosome 8p deletions and 8q gains are associated with tumor progression and poor prognosis in prostate cancer. Clin Cancer Res 16(1):56–64

Engellau J, Akerman M, Anderson H, Domanski HA, Rambech E, Alvegard TA et al (2001) Tissue microarray technique in soft tissue sarcoma: immunohistochemical Ki-67 expression in malignant fibrous histiocytoma. Appl Immunohistochem Mol Morphol 9(4):358–363

Erez A, Perelman M, Hewitt SM, Cojacaru G, Goldberg I, Shahar I et al (2004) Sil overexpression in lung cancer characterizes tumors with increased mitotic activity. Oncogene 23(31):5371–5377

Fejzo MS, Slamon DJ (2010) Tissue microarrays from frozen tissues-OCT technique. Methods Mol Biol 664:73–80

Fernebro E, Dictor M, Bendahl PO, Ferno M, Nilbert M (2002) Evaluation of the tissue microarray technique for immunohistochemical analysis in rectal cancer. Arch Pathol Lab Med 126(6):702–705

Fleischmann A, Schlomm T, Huland H, Kollermann J, Simon P, Mirlacher M et al (2008) Distinct subcellular expression patterns of neutral endopeptidase (CD10) in prostate cancer predict diverging clinical courses in surgically treated patients. Clin Cancer Res 14(23):7838–7842

Fleischmann A, Schlomm T, Kollermann J, Sekulic N, Huland H, Mirlacher M et al (2009) Immunological microenvironment in prostate cancer: high mast cell densities are associated with favorable tumor characteristics and good prognosis. Prostate 69(9):976–981

Gancberg D, Di Leo A, Rouas G, Jarvinen T, Verhest A, Isola J et al (2002) Reliability of the tissue microarray based FISH for evaluation of the HER-2 oncogene in breast carcinoma. J Clin Pathol 55(4):315–317

Geiszt M, Lekstrom K, Brenner S, Hewitt SM, Dana R, Malech HL et al (2003) NAD(P)H oxidase 1, a product of differentiated colon epithelial cells, can partially replace glycoprotein 91phox in the regulated production of superoxide by phagocytes. J Immunol 171(1):299–306

Ginestier C, Charafe-Jauffret E, Bertucci F, Eisinger F, Geneix J, Bechlian D et al (2002) Distinct and complementary information provided by use of tissue and DNA microarrays in the study of breast tumor markers. Am J Pathol 161(4):1223–1233

Hedvat CV, Hegde A, Chaganti RS, Chen B, Qin J, Filippa DA et al (2002) Application of tissue microarray technology to the study of non-Hodgkin's and Hodgkin's lymphoma. Hum Pathol 33(10):968–974

Hendriks Y, Franken P, Dierssen JW, De Leeuw W, Wijnen J, Dreef E et al (2003) Conventional and tissue microarray immunohistochemical expression analysis of mismatch repair in hereditary colorectal tumors. Am J Pathol 162(2):469–477

Hoos A, Stojadinovic A, Mastorides S, Urist MJ, Polsky D, Di Como CJ et al (2001a) High Ki-67 proliferative index predicts disease specific survival in patients with high-risk soft tissue sarcomas. Cancer 92(4):869–874

Hoos A, Urist MJ, Stojadinovic A, Mastorides S, Dudas ME, Leung DH et al (2001b) Validation of tissue microarrays for immunohistochemical profiling of cancer specimens using the example of human fibroblastic tumors. Am J Pathol 158(4):1245–1251

Hoos A, Stojadinovic A, Singh B, Dudas ME, Leung DH, Shaha AR et al (2002) Clinical significance of molecular expression profiles of Hurthle cell tumors of the thyroid gland analyzed via tissue microarrays. Am J Pathol 160(1):175–183

Howat WJ, Wilson SJ (2010) Resin technologies: construction and staining of resin TMA's. Methods Mol Biol 664:63–72

Howat WJ, Warford A, Mitchell JN, Clarke KF, Conquer JS, McCafferty J (2005) Resin tissue microarrays: a universal format for immunohistochemistry. J Histochem Cytochem 53(10):1189–1197

Jhavar S, Corbishley CM, Dearnaley D, Fisher C, Falconer A, Parker C et al (2005) Construction of tissue microarrays from prostate needle biopsy specimens. Br J Cancer 93(4):478–482

Kallioniemi A (2010) DNA copy number analysis on tissue microarrays. Methods Mol Biol 664:127–134

Kellner A, Matschke J, Bernreuther C, Moch H, Ferrer I, Glatzel M (2009) Autoantibodies against beta-amyloid are common in Alzheimer's disease and help control plaque burden. Ann Neurol 65(1):24–31

Khalique L, Ayhan A, Weale ME, Jacobs IJ, Ramus SJ, Gayther SA (2007) Genetic intra-tumour heterogeneity in epithelial ovarian cancer and its implications for molecular diagnosis of tumours. J Pathol 211(3):286–295

Kononen J, Bubendorf L, Kallioniemi A, Barlund M, Schraml P, Leighton S et al (1998) Tissue microarrays for high-throughput molecular profiling of tumor specimens. Nat Med 4(7):844–847

Kristiansen G, Fritzsche FR, Wassermann K, Jager C, Tolls A, Lein M et al (2008) GOLPH2 protein expression as a novel tissue biomarker for prostate cancer: implications for tissue-based diagnostics. Br J Cancer 99(6):939–948

LeBaron MJ, Crismon HR, Utama FE, Neilson LM, Sultan AS, Johnson KJ et al (2005) Ultrahigh density microarrays of solid samples. Nat Methods 2(7):511–513

McCabe A, Dolled-Filhart M, Camp RL, Rimm DL (2005) Automated quantitative analysis (AQUA) of in situ protein expression, antibody concentration, and prognosis. J Natl Cancer Inst 97(24):1808–1815

McCarthy F, Fletcher A, Dennis N, Cummings C, O'Donnell H, Clark J et al (2009) An improved method for constructing tissue microarrays from prostate needle biopsy specimens. J Clin Pathol 62(8):694–698

Minner S, Jessen B, Stiedenroth L, Burandt E, Kollermann J, Mirlacher M et al (2010) Low level HER2 overexpression is associated with rapid tumor cell proliferation and poor prognosis in prostate cancer. Clin Cancer Res 16(5):1553–1560

Mirlacher M, Simon R (2010) Recipient block TMA technique. Methods Mol Biol 664:37–44
Moch H, Schraml P, Bubendorf L, Mirlacher M, Kononen J, Gasser T et al (1999) High-throughput tissue microarray analysis to evaluate genes uncovered by cDNA microarray screening in renal cell carcinoma. Am J Pathol 154(4):981–986
Montgomery K, Zhao S, van de Rijn M, Natkunam Y (2005) A novel method for making "tissue" microarrays from small numbers of suspension cells. Appl Immunohistochem Mol Morphol 13(1):80–84
Nocito A, Bubendorf L, Tinner EM, Suess K, Wagner U, Forster T et al (2001) Microarrays of bladder cancer tissue are highly representative of proliferation index and histological grade. J Pathol 194(3):349–357
Oeggerli M, Tomovska S, Schraml P, Calvano-Forte D, Schafroth S, Simon R et al (2004) E2F3 amplification and overexpression is associated with invasive tumor growth and rapid tumor cell proliferation in urinary bladder cancer. Oncogene 23(33):5616–5623
Pires AR, de Souza SR (2010) Hypodermic needle without recipient paraffin block technique. Methods Mol Biol 664:53–61
Pires AR, Andreiuolo Fda M, de Souza SR (2006) TMA for all: a new method for the construction of tissue microarrays without recipient paraffin block using custom-built needles. Diagn Pathol 1:14
Press MF, Sauter G, Bernstein L, Villalobos IE, Mirlacher M, Zhou JY et al (2005) Diagnostic evaluation of HER-2 as a molecular target: an assessment of accuracy and reproducibility of laboratory testing in large, prospective, randomized clinical trials. Clin Cancer Res 11(18):6598–6607
Rimm DL, Camp RL, Charette LA, Costa J, Olsen DA, Reiss M (2001) Tissue microarray: a new technology for amplification of tissue resources. Cancer J 7(1):24–31
Robanus-Maandag EC, Bosch CA, Kristel PM, Hart AA, Faneyte IF, Nederlof PM et al (2003) Association of C-MYC amplification with progression from the in situ to the invasive stage in C-MYC-amplified breast carcinomas. J Pathol 201(1):75–82
Rubin MA, Dunn R, Strawderman M, Pienta KJ (2002) Tissue microarray sampling strategy for prostate cancer biomarker analysis. Am J Surg Pathol 26(3):312–319
Rui H, LeBaron MJ (2005) Creating tissue microarrays by cutting-edge matrix assembly. Expert Rev Med Devices 2(6):673–680
Sauter G (2010) Representativity of TMA studies. Methods Mol Biol 664:27–35
Sauter G, Mihatsch MJ (1998) Pussycats and baby tigers: non-invasive (pTa) and minimally invasive (pT1) bladder carcinomas are not the same! J Pathol 185(4):339–341
Schlomm T, Kirstein P, Iwers L, Daniel B, Steuber T, Walz J et al (2007) Clinical significance of epidermal growth factor receptor protein overexpression and gene copy number gains in prostate cancer. Clin Cancer Res 13(22 Pt 1):6579–6584
Schlomm T, Iwers L, Kirstein P, Jessen B, Kollermann J, Minner S et al (2008) Clinical significance of p53 alterations in surgically treated prostate cancers. Mod Pathol 21(11):1371–1379
Schoenberg Fejzo M, Slamon DJ (2001) Frozen tumor tissue microarray technology for analysis of tumor RNA, DNA, and proteins. Am J Pathol 159(5):1645–1650
Schraml P, Kononen J, Bubendorf L, Moch H, Bissig H, Nocito A et al (1999) Tissue microarrays for gene amplification surveys in many different tumor types. Clin Cancer Res 5(8):1966–1975
Shi S-R, Cote RJ, Taylor CR (2001) Antigen retrieval techniques: current perspectives. J Histochem Cytochem 49(8):931–938
Simon R, Burger H, Brinkschmidt C, Bocker W, Hertle L, Terpe HJ (1998) Chromosomal aberrations associated with invasion in papillary superficial bladder cancer. J Pathol 185(4):345–351
Simon R, Eltze E, Schafer KL, Burger H, Semjonow A, Hertle L et al (2001a) Cytogenetic analysis of multifocal bladder cancer supports a monoclonal origin and intraepithelial spread of tumor cells. Cancer Res 61(1):355–362
Simon R, Nocito A, Hubscher T, Bucher C, Torhorst J, Schraml P et al (2001b) Patterns of her-2/neu amplification and overexpression in primary and metastatic breast cancer. J Natl Cancer Inst 93(15):1141–1146

Simon R, Richter J, Wagner U, Fijan A, Bruderer J, Schmid U et al (2001c) High-throughput tissue microarray analysis of 3p25 (RAF1) and 8p12 (FGFR1) copy number alterations in urinary bladder cancer. Cancer Res 61(11):4514–4519

Simon R, Struckmann K, Schraml P, Wagner U, Forster T, Moch H et al (2002) Amplification pattern of 12q13-q15 genes (MDM2, CDK4, GLI) in urinary bladder cancer. Oncogene 21(16):2476–2483

Simon R, Atefy R, Wagner U, Forster T, Fijan A, Bruderer J et al (2003) HER-2 and TOP2A coamplification in urinary bladder cancer. Int J Cancer 107(5):764–772

Tapia C, Glatz K, Novotny H, Lugli A, Horcic M, Seemayer CA et al (2007) Close association between HER-2 amplification and overexpression in human tumors of non-breast origin. Mod Pathol 20(2):192–198

Tennstedt P, Köster P, Brüchmann A, Mirlacher M, Haese A, Steuber T et al (2012) The impact of the number of cores on tissue microarray studies investigating prostate cancer biomarkers. Int J Oncol 40(1):261–268

Torhorst J, Bucher C, Kononen J, Haas P, Zuber M, Kochli OR et al (2001) Tissue microarrays for rapid linking of molecular changes to clinical endpoints. Am J Pathol 159(6):2249–2256

Tzankov A, Zimpfer A, Lugli A, Krugmann J, Went P, Schraml P et al (2003) High-throughput tissue microarray analysis of G1-cyclin alterations in classical Hodgkin's lymphoma indicates overexpression of cyclin E1. J Pathol 199(2):201–207

Veeck J, Dahl E (2010) RNA expression analysis on formalin-fixed paraffin-embedded tissues in TMA format by RNA in situ hybridization. Methods Mol Biol 664:135–150

Went PT, Dirnhofer S, Bundi M, Mirlacher M, Schraml P, Mangialaio S et al (2004) Prevalence of KIT expression in human tumors. J Clin Oncol 22(22):4514–4522

Zhang D, Salto-Tellez M, Putti TC, Do E, Koay ES (2003) Reliability of tissue microarrays in detecting protein expression and gene amplification in breast cancer. Mod Pathol 16(1):79–84

Zhao S, Natkunam Y (2010) Building "tissue" microarrays from suspension cells. Methods Mol Biol 664:93–101

Chapter 10
Recent and Future Developments of Microarrays: Miniaturization and Lab-on-Chip Approaches

Inbal Tsarfati-BarAd and Levi A. Gheber

10.1 Introduction

The last two decades have seen an impressive development in the field of microarrays, which have grown into an essential tool in genomics, proteomics, and diagnostic research. The foundation of the microarray technology was set with Ekins' "Multi-analyte immunoassay" in the 1980s (Ekins 1989). The need for a global tool that can process large amounts of information in a short time led to the development of the DNA chip (Schena et al. 1995) and subsequently the protein chip (MacBeath and Schreiber 2000). Presently, microarrays of DNA, RNA, antibodies, cell lysates, purified proteins, tissue sections, and embedded cells are used for analysis in a variety of fields (Joos and Kroeger 2008).

All microarrays ("biochips") share a similar format: capture molecules are immobilized on a solid surface as a rectangular array of small circular areas (termed "spots"). Typically, various spots consist of different capture molecules intended to bind various target molecules. The array is then exposed to the tested sample and queried for the presence of the target molecules (that bind specifically the capture molecules). This can be done in a number of ways, such as using a labeled target molecule (a "direct" assay), using a secondary, labeled, target molecule that detects the target bound to the primary capture molecule (a "sandwich" assay), or using a "binding competition" assay, where different concentrations of the (labeled) secondary capture molecule are mixed with the sample and subsequently the mixture is applied to the array; only the free target molecule (that was not bound by the secondary capture molecule) binds the immobilized capture molecule. The label is needed in order to report on the binding event and is typically a fluorescent

I. Tsarfati-BarAd • L.A. Gheber (✉)
Avram and Stella Goldstein-Goren Department of Biotechnology Engineering, The Ilse Katz Institute for Nanoscale Science and Technology, Ben-Gurion University of the Negev, Beer-Sheva, Israel
e-mail: glevi@bgu.ac.il

B. Jordan (ed.), *Microarrays in Diagnostics and Biomarker Development*,
DOI 10.1007/978-3-642-28203-4_10, © Springer-Verlag Berlin Heidelberg 2012

Fig. 10.1 An example of a direct (*left*) and indirect (*right*) assay

molecule or an enzyme that catalyzes a chemiluminescent reaction. The light signal is then collected and quantified with appropriate optics, detectors, and software (Liotta et al. 2003) (see Fig. 10.1).

Current microarrays are manufactured on chemically modified glass substrates that can bind covalently or adsorb the capture biomolecules. The typical diameter of spots is roughly 100 μm, and the center-to-center spacing is around 300 μm (Gheber 2007). The majority of target molecules are fluorescently labeled, since fluorescence detection leads to the highest sensitivity of quantization (Amersham-Biosciences 2002). This microarray structure requires up to five different components in order to yield a full array: the chip's substrate, a chemically modified surface; the spotter, a device that applies the patterned spots on the surface; a liquid handling system for the blocking, washing, and hybridization; a scanner, an automated machine for the reading of a large number of spots; and software for the analysis of the results. These are in addition to the equipment needed for the production, purification, and/or labeling of the array's biomolecules (Blohm and Guiseppi-Elie 2001). In addition, trained personnel handling all the operations and equipment are a must. Thus, contemporary technology limits the use of these high-throughput and sensitive tools to large research laboratories, major pharmaceutical companies, or advanced hospitals (Gheber 2007).

To enable widespread use of the highly parallel assay concept, both for research and even more importantly for diagnostic and monitoring applications, researchers aim to make microarrays portable, stable, cheap, and fast. This challenge requires a better understanding of the limitations of the current technology and a careful analysis of the approaches to solve the problems. The roadmap leading to such breakthrough seems to have a single enabling challenge at its start: miniaturization of the active site—the "spot." Miniaturization in the "making it small" sense has already been achieved, following over a decade of development of nanobiolithography techniques. It is the "making it small and useful" sense that proves to be problematic. Miniaturization of spots leads to tremendous loss in signal-to-noise ratio (SNR). To avoid this, new immobilization surfaces need to be developed. New light collection strategies have to be devised, consistent with the surfaces, and signal enhancement methods (e.g., based on plasmonics) need to be perfected. Importantly, all the above have been and are being developed on separate tracks. The need for a harmonious, simultaneous approach cannot be overstated.

In the following, we discuss each key factor hindering the development of portable biochips and emphasize the interconnection between all of them.

10.2 Key Factors Hindering Development of Portable Arrayed Biochips

10.2.1 Miniaturization of Spot Size

The size of spots turns out to be a major obstacle in the way towards portability of the microarray technology (Tsarfati-BarAd et al. 2011). The fluorescent spots emit low levels of light, and highly efficient light collection is required to detect and quantify the signal. In order to achieve this, objectives with high numerical aperture (NA) are used. These, in turn, have large magnification and a small field of view that can image only a few spots (with dimensions and spacing as discussed above) at once. To image the whole array, an automated array scanner is used—a heavy and expensive machine. Drastically reducing the spot size should render the scanner unnecessary since the whole array would be visible in the view field of the objective. This would constitute a significant step towards microarray portability.

Another aspect of spot size that should be considered is the biochip's cost. Biomolecules require a long, complex, and expensive process to produce and purify. This normally leads to a disproportionately high contribution of biomolecule cost to the total price of a biochip. By reducing the spot size by, e.g., 100-fold, the amount of biomaterial needed will be reduced by 10^4, a substantial saving in the microarray production cost.

The reduction of the spot size can also contribute to reducing the time-to-result of assays. This is so since mass and heat transfer distances are significantly reduced.

Yet another concept consistent with the need to reduce spot size is the "ambient analyte theory" (Ekins and Chu 1991), showing that given a small enough spot with a maximal binding site density, the capture molecule concentration and sample volume will not affect the ligand occupancy. If these conditions are not met, the target molecules captured by the capture molecules represent in fact a significant fraction of all the target molecules in the sample, thus effectively reducing the target concentration; the measurement itself changes the result of the measurement. Smaller spots, though, enable better sensitivity and the use of smaller sample volume.

10.2.2 Label-Free Detection

Labeling plays a critical role in microarray sensitivity and dynamic range. Fluorescent labeling methods have been developed over the years, and fluorescent dyes have become brighter and more resistant to photobleaching, thus improving the array sensitivity and stability over time (Amersham-Biosciences 2002). However, labeling exhibits a number of drawbacks.

The chemical labeling might disrupt the molecule structure, changing the binding affinity or even blocking the molecule's binding site. Moreover, molecules (especially small ones) may not necessarily have the functional group required for labeling. Even if these two issues are resolved, a fundamental problem is that the need to label molecules prevents real-time detection. Labeling the target molecule or applying a secondary labeled molecule is not compatible with a device that continuously monitors, e.g., pathogens in the water of a river, bacteria in the atmosphere of a hospital, or biological threats in the stream of beverage just before bottling in the factory. A variety of label-free detection techniques were thus developed over the years. The most widely used are surface plasmon resonance (SPR), quartz crystal microbalance (QCM), microcantilevers, and AFM-based methods, to name a few. Details are beyond the scope of this chapter; however, for recent reviews, see Chandra et al. (2011) and Yu et al. (2006). A general remark on currently used label-free detection methods is that they lack the ability to address small areas and are thus not compatible with microarrays (and certainly not with miniaturized arrays).

10.2.3 Integration

Assays on current microarrays are carried out using separate instruments and facilities: blocking, incubation with the sample, and washing are performed manually; the arrays are then imaged using a scanner; and the data are analyzed later on a (dedicated) computer. All these require a well-equipped laboratory with well-trained personnel. These components require miniaturization and integration on a portable, stable, fast, stand-alone device: lab on chip (LOC) (Craighead 2006). Not only the active sites of the microarray need to be much smaller, but the sample handling systems, light sources, light collection, detectors, and power sources need to be miniaturized and integrated on the same platform. One additional aspect of integration that tends to be overlooked is the fact that such a platform would require communication abilities. A LOC system, as described above, that collects the sample, handles it, and detects and quantifies the signal, once deployed in the "field," would require the visit of a human to record the readings, unless it can remotely communicate with them. If such visits are required, the LOC approach does not present significant advantages over the collection (by a human) of the sample in the field and its analysis in the laboratory.

10.3 Recent Advances in the Microarray Field

10.3.1 Miniaturization of Spots

Traditional nanolithography approaches imported from nanoelectronics are not compatible with biomolecules, since they typically use extreme conditions such as high vacuum, strong acids, and elevated temperatures. The last 10–15 years have

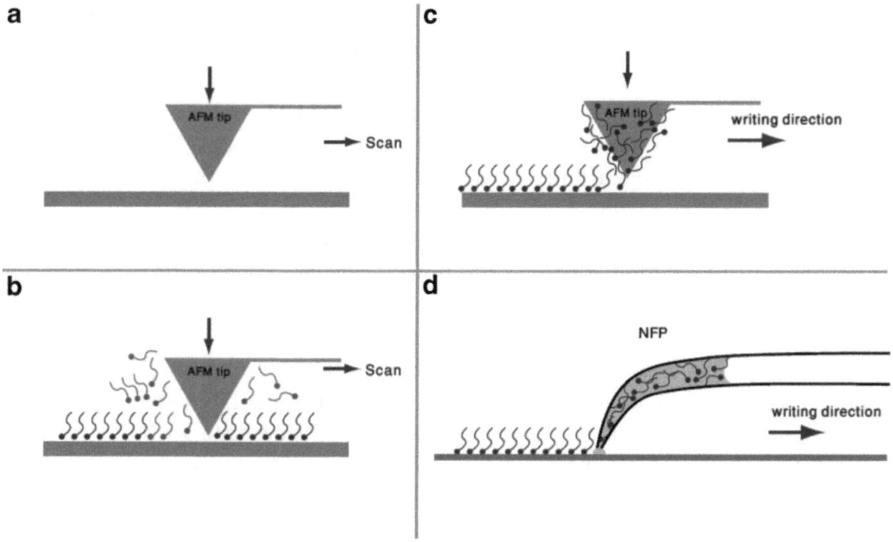

Fig. 10.2 Scanning probe microscopy-based methods for nanobiolithography: (**a**) an AFM tip. (**b**) Nanografting—AFM tip is used to "shave" a self-assembled monolayer serving as mask, allowing molecules in the solution to bind to the shaved surface. (**c**) Dip-pen nanolithography—an AFM tip is dipped in a solution, and subsequently, when the tip touches the surface, capillary forces cause the molecules to transport from the tip to the substrate. (**d**) Nanofountain pen—a glass nanopipette, filled with a solution, is replacing the AFM tip

witnessed the development of a myriad of methods for fabrication of biological nanostructures. The most promising ones are divided in two main groups: scanning probe microscopy (SPM)-based methods and nanoimprint lithography (NIL)-based methods.

10.3.1.1 SPM-Based Nanolithography

One natural candidate for nanobiolithography is the atomic force microscope (AFM) (see Fig. 10.2a), since it operates under ambient conditions and possesses high positioning precision. The evolution of SPM-based methods is briefly described below.

Nanografting—In nanografting, the AFM tip is used to "shave" a self-assembled monolayer that serves as a mask (see Fig. 10.2b), and the substrate is then exposed to the adsorption of other molecules from the surrounding solution (Wadu-Mesthrige et al. 1999). Bano and coworkers utilized nanografting lithography to fabricate protein arrays of various sub-μm sizes. Ethylene glycol-modified thiol monolayers on gold were grafted and replaced by ssDNA on which DNA–protein conjugates were immobilized. In order to prove protein functionality, some DNA–protein conjugates such as DNA–streptavidin conjugates were immobilized to the surface and probed with anti-streptavidin IgG (Bano et al. 2009). It should be noted that the binding was only shown with AFM scanning and height comparison; nevertheless, this assay indicates the feasibility of nanografting as a tool for the fabrication of nanobioarrays.

Dip-Pen Nanolithography (*DPN*)—In DPN, an AFM probe is used as a pen dipped in a solution ("ink") consisting of biological molecules. These biological molecules have chemical affinity for a solid-state substrate so that capillary forces cause the molecules to transport from the tip to the substrate (Piner et al. 1999; Salaita et al. 2007) (see Fig. 10.2c). This method allows one to directly apply very small amounts of solution at precise positions. However, most of the time, both the AFM tip and the substrate need to be chemically modified for the solution to flow smoothly.

Mirkin's group used DPN to fabricate both DNA microarrays (Demers et al. 2002) and antibody microarrays (Vega et al. 2006). For the DNA microarrays, the AFM tips were modified with 3-aminopropyltrimethoxysilane to promote DNA adhesion on the tip. Then the DNA "ink" was deposited on silicon oxide wafers. The hybridization of about 1-μm spots was imaged using fluorescent labeling. For the immunoarray, DPN was used to pattern 16-mercaptohexadecanoic acid (MHA) on gold thin-film substrate, while the rest of the substrate was passivated using with 11-mercaptoundecylpenta(ethylene glycol), to prevent nonspecific binding. The MHA groups were coordinated with Zn ions which were demonstrated to immobilize the antibodies. This assay was used to demonstrate a sandwich assay and fluorescently report on the presence of influenza virus particles.

Nanofountain Pen—A nanofountain pen is a glass cantilevered nanopipette that is filled with a solution, which is drawn by capillary forces to the tapered tip of the pipette. Surface tension prevents the liquid from flowing out until contact with the surface is made (Taha et al. 2003) (see Fig. 10.2d). This method is able to fabricate protein nanostructures without additional preparations of the NFP or the surface. Bruckbauer et al. (2004) used NFP to fill antibody solution in small holes etched on Au-coated glass, fabricated using focused ion beam (FIB). Labeled IgG anti-IgG assay was imaged in order to prove biological activity (Bruckbauer et al. 2004).

10.3.1.2 Imprint Lithography

Imprint lithography is a general name for methods exploiting a structured mold or a template, to produce a pattern (Truskett and Watts 2006) (see Fig. 10.3). Nanoimprint lithography techniques usually rely on nanostructuring of polymers that later serve as molds. Manufacture of molds is usually done using one of two methods. *Conventional photolithography* (Fig. 10.3a) uses a photoresist layer exposed to UV light through a mask. Next, the substrate is exposed to a developing solution that can etch either the exposed parts (positive photolithography) or the unexposed parts (negative photolithography). *Molding and embossing processes* (Fig. 10.3b) push a rigid mold into a thin thermoplastic layer spun on a substrate, heated above its glass transition temperature. After the polymer cools off, the mold

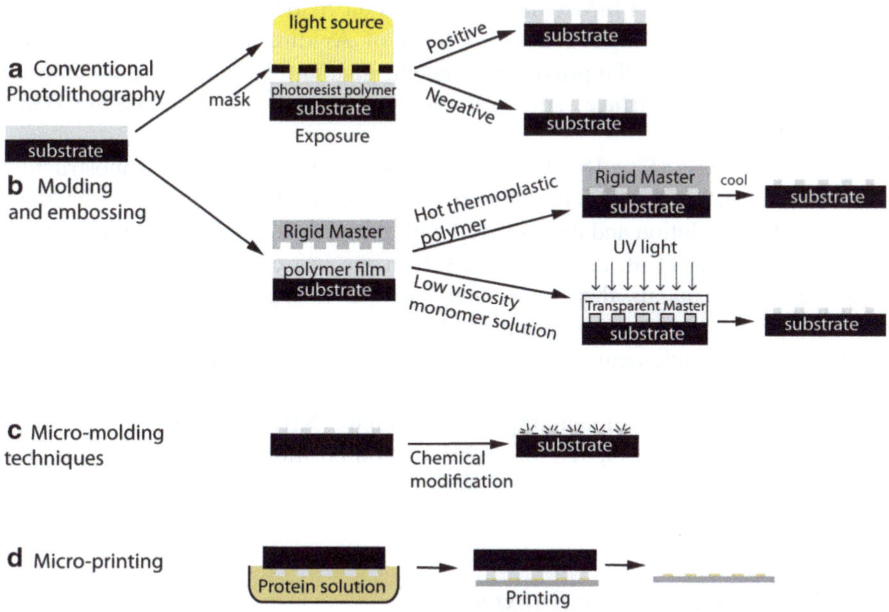

Fig. 10.3 Imprint lithography. (**a**) Conventional photolithography—exposing photoresist layer to UV light trough a mask and developing either the exposed parts (positive photolithography) or the unexposed parts (negative photolithography). (**b**) Molding and embossing—a rigid master is pushed into a polymer layer that is stiffened according to the pattern. (**c**) Micromolding—the polymer pattern is chemically modified to further bind chemical or biological molecules. (**d**) Contact printing—the polymer pattern is dipped into a solution and applied with force on a surface (like a "stamp")

is separated and the structure remains imprinted in the polymer. Instead of a thermoplastic layer, one can use in the first stage a low-viscosity monomer that is either spin coated (UV-nanoimprint lithography) or applied as small droplets (step and flash imprint lithography). In the second step, the solution is UV-polymerized in contact with the mold, and the pattern is thus transferred to the polymer. After the polymer nanostructures are fabricated, they are typically used in one of two ways described below (Truskett and Watts 2006).

Micromolding Techniques (Fig. 10.3c)—In micromolding, the polymer nanostructure undergoes chemical modification so it will be able to attach biomolecules. Moorcroft et al. (2005) oxidized PDMS nanostructures and altered them using (1) epoxy silane (3-glycidoxypropyltrimethoxysilane) and (2) PEG (polyethylene glycol). DNA was synthesized on the PEG residues with a DNA synthesizer. The DNA was then hybridized with complementary fluorescently labeled DNA (Moorcroft et al. 2005). Hoff et al. imprinted a silicon mold on poly (methyl methacrylate) (PMMA) spin coated on a thin layer of silicon oxide. After removing the silicon mold, the exposed silicon oxide was etched, and subsequently the PMMA was removed, so that a silicon oxide pattern remained in the region covered by the PMMA mold. The silicon oxide was then modified with chemical and biological reagents to a layer of Avidin, in order to immobilize biotin labeled proteins. To prove the concept, goat anti-catalase antibodies were immobilized to the surface, and captured fluorescently labeled catalase (Hoff et al. 2004).

Contact Printing (Fig. 10.3d)—In contact printing, the polymer nanostructure (usually an elastomer such as polydimethylsiloxane [PDMS]) is used as a "stamp." It is dipped in a solution and then applied with pressure on a substrate (Truskett and Watts 2006). Xiao et al. reported in 2002 on the use of PDMS stamps to print structures of chemicals that were then used as a source for oligonucleotides synthesis. The synthesized oligonucleotides were then hybridized with fluorescently labeled complementary strands to show microarray feasibility (Xiao et al. 2002).

One common feature of nanoimprint lithography (NIL) methods is the difficulty of fabricating features (spots) of different compositions in the same printing session.

10.3.2 Immobilization Substrates

Microarray sensitivity, reproducibility, and dynamic range are derived from the substrate properties: linking chemistry, binding site density, homogeneity, and background noise produced either by the surface itself or nonspecific binding on the surface (Blohm and Guiseppi-Elie 2001). The microarray industry typically uses glass surfaces as substrates, because of their low intrinsic background. The search for higher signal and lower noise has driven surface manufacturers to develop high-density and uniform coating on microscope glass substrates (Amersham-Biosciences 2002).

Most of the nanofabrication methods described above are based on substrates such as gold and silicon. These substrates exhibit clear advantages for the fabrication of nanostructures since they are homogeneous, very flat, and extremely well characterized. However, these substrates are not optimal for microarrays and specifically for fluorescence-based detection. In addition to their high price, these surfaces strongly reflect incident light, thus creating high background; they are not transparent, so they restrict readout methods; and in the case of gold, they might strongly quench the fluorescence, thus reducing signal intensity. The choice of glass as the standard substrate of conventional microarrays is based precisely on these considerations, which are completely sacrificed in most nanolithography methods, in favor of the feature size. Moreover, the substrates for nanolithography are modified with chemicals (Stoll et al. 2002), which are not necessarily optimized for lowest possible autofluorescence.

The challenge is thus to fabricate dramatically smaller bioarrays on relevant substrates, optimized for best signal-to-noise ratio (SNR), rather than on substrates that are optimized for size of spot (but hamper SNR). While this is well understood,

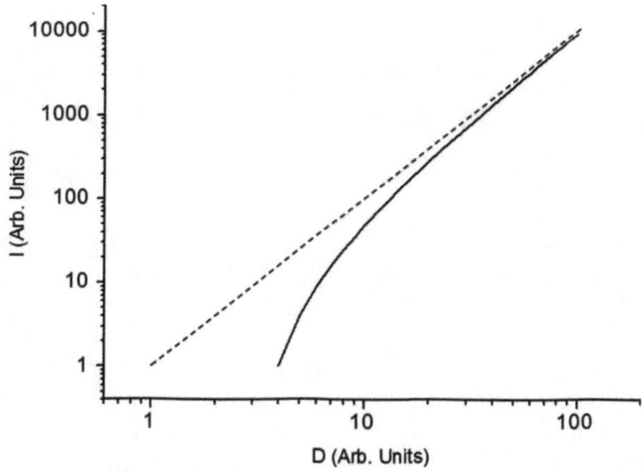

Fig. 10.4 Dependence of the fluorescence intensity on the spot diameter (log–log plot). For an ideal surface, intensity is proportional to the squared diameter: $I \sim (D)^2$ (*dashed line*). According to the model presented in Tsarfati-BarAd, Sauer, et al. (2011), on a realistic surface, the intensity decreases faster like $I \sim (D - \mathrm{b})^2$, where b is the average distance between binding sites, and vanishes for $D = b$. In this example, $b = 3$ (*continuous line*) (Reprinted with permissions from Tsarfati-BarAd, Sauer et al. 2011)

no examples of such achievements have been reported. Typically researchers report a dramatic loss in SNR upon reduction of spot size and conclude that it is impractical.

We have recently shown that when reducing spot size, the density of binding sites on a real surface is the key factor for the loss in SNR (and not the background of the substrates, which is very low to begin with). The intensity of the signal of a spot is proportional only to the *binding* area, rather than the whole area covered by the drop. When the size of the printed spot approaches the dimensions of the nonbinding gap between binding areas, there is a significant decrease in the number of molecules bound and consequently a decrease in the signal, much beyond the one naively predicted on an ideal surface (Tsarfati-BarAd et al. 2011) (see Fig. 10.4). Thus, when miniaturizing the array, it is crucial to assess which substrates are suitable for nanopatterning. We have developed a mathematical model that predicts the loss of signal upon reduction of the spot diameter, as a function of the density of binding sites. The model suggests a way to measure the "*b-value*," a parameter representative of the density of binding sites, and to compare it between various substrates. NFP was used to deposit spots of IgG of varying diameters on different commercial substrates. Following a direct assay, fluorescence from an array of ~150 spots was captured in one view field of a standard fluorescence microscope. We further quantified the fluorescence intensity of each spot and its diameter. This allowed confirming the dependence of signal intensity on the spot diameter predicted by the model. Indeed, we have been able to demonstrate ~1-μm spots performing successfully on glass-based, standard surfaces, which have been correctly chosen (see Fig. 10.5).

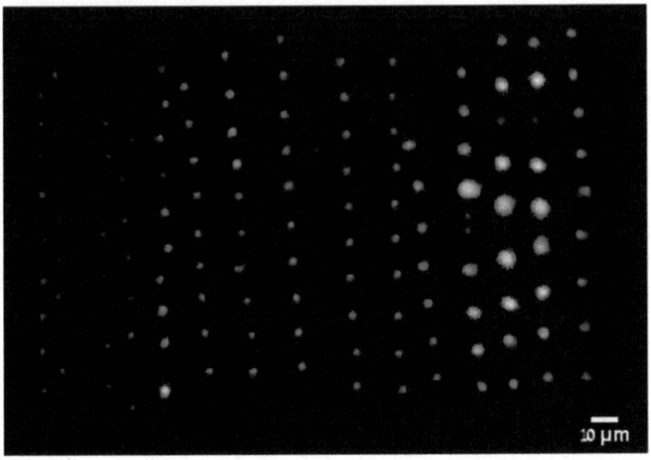

Fig. 10.5 Fluorescence image of miniaturized immunoassay on hydrogel-coated glass substrate. Goat anti-rabbit IgG spots were printed with various diameters using a nanofountain pen and probed with fluorescently labeled rabbit IgG. Spot diameters start at 10 µm and down to ~200 nm (Reprinted with permissions from Tsarfati-BarAd, Sauer et al. 2011)

Another antibody assay on glass substrates was fabricated using nanocontact printing. Stripes of secondary antibodies were probed with labeled antibodies (Graber et al. 2003). These are first and important steps in combining nanolithography methods with microarray technology for the future fabrication of nanoarrays.

10.3.3 Increasing Signal-to-Noise Ratio

Increasing SNR can be achieved by increasing the signal, reducing the noise (background), and (preferably) a combination. A number of approaches to achieving increase in SNR deserve special discussion.

10.3.3.1 Protein Orientation

The layer of capture molecules plays, in fact, the role of an immobilization layer for the target molecules. This "second floor" of the multistory building has too a finite density of binding sites. Obviously, this density is primarily dictated by the density of the immobilization layer discussed above. However, additional reduction of binding site density is due to the immobilization process that might induce conformation changes and even interfere with the active site in the capture molecules. This is especially evident with IgG antibodies, which are "Y"-shaped

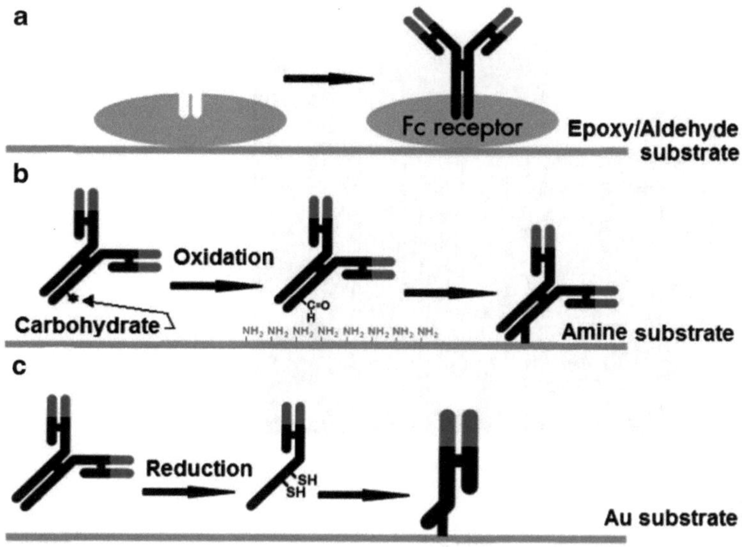

Fig. 10.6 Antibody orientation methods. (**a**) Binding through Fc receptors such as protein A, protein G, or recombinant protein A/G. (**b**) Oxidation of carbohydrate moieties on the Fc region of the antibodies. The resulting aldehyde groups are attached to amino slide. (**c**) Reduction of antibodies into Fab-SH fragments and attachment to an amine-coated glass surface, modified to attach thiol residues

asymmetric molecules. Many methods of protein immobilization use primary amines; for IgG, these occur on lysine residues and the N-terminus of each polypeptide chain. They are numerous and distributed over the entire antibody. The immobilization of IgG using this chemistry, thus, binds antibodies in random configurations, many of them unfavorable (unable to bind the target) (Kusnezow and Hoheisel 2003). Antibody orientation techniques were borrowed over the years from chromatography techniques, and three methods in particular have been reported (see Fig. 10.6): using Fc region receptors (such as protein A or G) (Fig. 10.6a); oxidizing the carbohydrate moieties on the Fc region and covalently binding them to amine-coated surfaces (Fig. 10.6b); and reduction of the disulfide bonds in the Fc region and attachment to gold surfaces through the thiol residues (Fig. 10.6c).

All of these methods have their merits and weaknesses; however, only a few comprehensive studies have been carried out that fully assess the superiority of one method over the others. One example is reported in Vijayendran and Leckband (2001). This field needs to be further studied to better understand the binding kinetics and appropriately choose the best immobilization method for each assay. Successfully immobilizing capture proteins in a favorable orientation may significantly increase SNR to offset the signal loss due to miniaturization of spots.

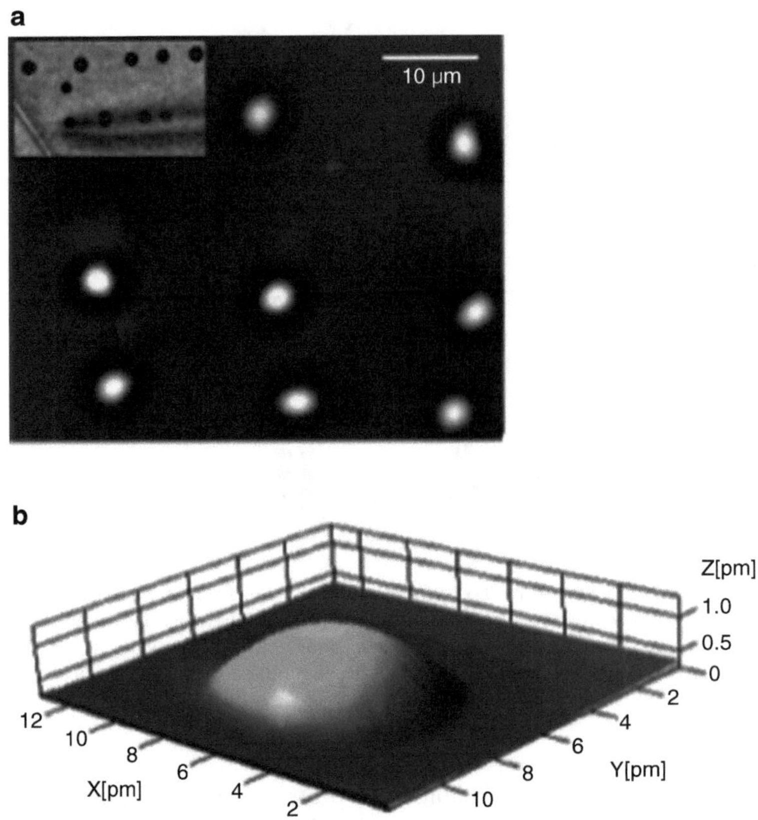

Fig. 10.7 (**a**) Bright-field image of an array of polymer microlenses manufactured with a nanofountain pen. The light source is behind the surface. (**b**) An AFM characterization of one microlens (Reprinted with permissions from Sokuler and Gheber 2006)

10.3.3.2 Microlenses

One approach to enhancing the signal is by increasing the collection efficiency of the emitted fluorescent signal. This can be done by the precise positioning of microlenses that collect the signal and concentrate it in their focal plane. We have demonstrated this concept (Sokuler and Gheber 2006; Sokuler et al. 2007) using an NFP-based technique for the manufacture of polymer microlenses. These microlenses can be fabricated on the other side of an array slide, facing the spots (see Fig. 10.7), an approach that is also consistent with integration of the reading system on the chip (no external objective, but rather an integrated, individual objective for each spot).

10.3.4 Integration

The concept of miniaturized (micro) total analysis systems (μ-TAS) has been around since 1990 (Manz et al. 1990), a concept with the potential to revolutionize diagnostics. Patient treatment requires fast, high-quality, near-patient information for optimal diagnosis. Miniaturized diagnostic instruments such as point-of-care devices (POC) and lab on chip (LOC) can also contribute to medicine paradigm shifting—from curative to predictive and preemptive (Duer et al. 2010). In addition to the need for miniaturization discussed previously, the integration of the various components forming the system onto a single platform that can perform in the field is a challenge to overcome.

10.3.4.1 Stability

A stand-alone analysis system needs to operate away from the lab, perhaps following storage, and preferably for long periods of time (continuous monitoring of a target molecule). For some applications, the use of biomaterials is challenging since they (especially proteins) are not stable out of their natural environment. Moreover, the production and purification of specific proteins are complex, expensive, and not always available. This is a reason to search for artificial substitutes. An example of such substitute is molecularly imprinted polymers (MIPs) that hold the promise to serve as "plastic antibodies" (Haupt and Mosbach 1998). A monomer mixture containing various functional groups and cross-linker is polymerized in the presence of a template molecule (the "antigen"). After polymerization, the template molecules are extracted, thus leaving cavities that are complementary to the template molecule and can specifically recognize it (Bui and Haupt 2010) (see Fig. 10.8). We have shown that NFP can be used to print miniature spots of MIPs (Belmont et al. 2007), and detect the binding of a fluorescent molecule (fluorescein); this demonstrates a way of using MIPs as the spots in a miniaturized array, while the repetitive binding–washing cycles show the usefulness of such stable materials.

Fig. 10.8 Molecularly imprinted polymers. (**a**) Monomers with functional groups, (**b**) cross-linker, and (**c**) template molecule are mixed (1). Next, the complex is polymerized (2) and the template is extracted (3), leaving a cavity that selectively recognizes and binds the template molecule (4) (Reprinted with permission from Haupt 2003)

10.3.4.2 Label-Free Detection

Significant steps towards integrating these approaches with label-free detection were reported (Kantarovich et al. 2009, 2010) where minuscule MIP spots have been arrayed on a surface-enhanced Raman spectroscopy (SERS) surface to detect and map the binding of a small, biologically relevant molecule (propranolol).

10.3.4.3 Integration with Microfluidic Platforms

Microfluidic devices are generated using a variety of microlithography methods similar to methods used in microelectronics. When combined with microarrays, these devices can be used to design compact diagnostic systems that analyze small volumes of the sample and decrease the cost of biomaterials. It is also an important step towards automated sample preparation and reduced sample processing time (Situma et al. 2006). When both the microfluidic chamber and the microarray area are small, the hybridization is more efficient—the probe molecules can move throughout the chamber and encounter every single probe (Wang and Li 2011).

However, the use of microfluidics for microarray applications may present new challenges for the construction of the device. The high surface-to-volume ratio characterizing microfluidics can increase nonspecific analyte/wall interactions. Integration with optical and electrical components needs to be developed. Manufacturing cost is still high, and less expensive methods must be developed (Situma et al. 2006).

10.4 Concluding Remarks

In the future years, the microarray is expected to become smaller, faster, and more sensitive. In doing so, the miniaturized microarray (whether the "nano" adjective will be justified by the real dimensions, or just a way of distinguishing it from its "micro" ancestor) will be gradually integrated at the heart of increasingly complex platforms. A variety of supporting subsystems, including sample handling, reading systems, remote transmission of data, and/or instructions, will have to be developed in a harmonious way, rather than on separate avenues. For example, the immobilization chemistry will need to be compatible with the materials from which the microfluidic liquid handling system is made (e.g., a miniaturized array manufactured directly in a microchannel) and with materials used for signal enhancement (e.g., metal nanostructures) in addition to wetting properties determining the size of the spots and low (fluorescence) background. Label-free detection methods will need to be compatible with miniaturization and integration into such small autonomous platforms, and light sources and detectors will need to be integrated on the platforms, as well as power sources.

Over the last years, there have been some important developments in these directions. However, much more work is required across the various fields, as well as integrating these into one cohesive technology, before a commercially viable solution will be achieved. It seems that due to the huge diversity of the scientific and technological disciplines coming into contact at this intersection, teams of scientists, with considerable overlapping abilities, will need to closely collaborate in attaining these goals. It is obvious that successful all-in-one platforms, with miniaturized bioarrays at their heart, will tremendously improve diagnostics and environmental monitoring, thus improving both human health and standard of living.

References

Amersham-Biosciences (2002) The microarray handbook. Amersham Biosciences, Buckinghamshire

Bano F, Fruk L et al (2009) Toward multiprotein nanoarrays using nanografting and DNA directed immobilization of proteins. Nano Letters 9(7):2614–2618

Belmont AS, Sokuler M et al (2007) Direct writing of molecularly imprinted microstructures using a nanofountain pen. Appl Phys Lett 90(19):193101-1–193101-3

Blohm DH, Guiseppi-Elie A (2001) New developments in microarray technology. Curr Opin Biotechnol 12(1):41–47

Bruckbauer A, Zhou D et al (2004) An addressable antibody nanoarray produced on a nanostructured surface. J Am Chem Soc 126(21):6508–6509

Bui BTS, Haupt K (2010) Molecularly imprinted polymers: synthetic receptors in bioanalysis. Anal Bioanal Chem 398(6):2481–2492

Chandra H, Reddy PJ et al (2011) Protein microarrays and novel detection platforms. Expert Rev Proteomics 8(1):61–79

Craighead H (2006) Future lab-on-a-chip technologies for interrogating individual molecules. Nature 442(7101):387–393

Demers LM, Ginger DS et al (2002) Direct patterning of modified oligonucleotides on metals and insulators by dip-pen nanolithography. Science 296(5574):1836–1838

Duer R, Lund R et al (2010) In-plane parallel scanning: a microarray technology for point-of-care testing. Anal Chem 82(21):8856–8865

Ekins RP (1989) Multi-analyte immunoassay. J Pharm Biomed Anal 7(2):155–168

Ekins RP, Chu FW (1991) Multianalyte microspot immunoassay—microanalytical compact-disk of the future. Clin Chem 37(11):1955–1967

Gheber LA (2007) Nanobiolithography of biochips. In: Marks RS, Cullen DC, Karube I, Lowe CR, Weetall HH (eds) Handbook of biosensors and biochips, vol 2. Wiley, Chichester, pp 771–783

Graber DJ, Zieziulewicz TJ et al (2003) Antigen binding specificity of antibodies patterned by microcontact printing. Langmuir 19(13):5431–5434

Haupt K (2003) Molecularly imprinted polymers: the next generation. Anal Chem 75(17):376A–383A

Haupt K, Mosbach K (1998) Plastic antibodies: developments and applications. Trends Biotechnol 16(11):468–475

Hoff JD, Cheng LJ et al (2004) Nanoscale protein patterning by imprint lithography. Nano Letters 4(5):853–857

Joos T, Kroeger P (2008) New frontiers in microarray technology development. Curr Opin Biotechnol 19(1):1–3

Kantarovich K, Tsarfati I et al (2009) Writing droplets of molecularly imprinted polymers by nano fountain pen and detecting their molecular interactions by surface-enhanced Raman scattering. Anal Chem 81(14):5686–5690
Kantarovich K, Tsarfati I et al (2010) Reading microdots of a molecularly imprinted polymer by surface-enhanced Raman spectroscopy. Biosensors Bioelectron 26(2):809–814
Kusnezow W, Hoheisel JD (2003) Solid supports for microarray immunoassays. J Mol Recogn 16(4):165–176
Liotta LA, Espina V et al (2003) Protein microarrays: meeting analytical challenges for clinical applications. Cancer Cell 3(4):317–325
MacBeath G, Schreiber SL (2000) Printing proteins as microarrays for high-throughput function determination. Science 289(5485):1760–1763
Manz A, Graber N et al (1990) Miniaturized total chemical-analysis systems—a novel concept for chemical sensing. Sens Actuators B Chem 1(6):244–248
Moorcroft MJ, Meuleman WRA et al (2005) In situ oligonucleotide synthesis on poly(dimethylsiloxane): a flexible substrate for microarray fabrication. Nucleic Acids Res 33(8):e75
Piner RD, Zhu J et al (1999) Dip-Pen nanolithography. Science 283(5402):661–663
Salaita K, Wang YH et al (2007) Applications of dip-pen nanolithography. Nat Nanotechnol 2(3):145–155
Schena M, Shalon D et al (1995) Quantitative monitoring of gene-expression patterns with a complementary-DNA microarray. Science 270(5235):467–470
Situma C, Hashimoto M et al (2006) Merging microfluidics with microarray-based bioassays. Biomol Eng 23(5):213–231
Sokuler M, Gheber LA (2006) Nano fountain pen manufacture of polymer lenses for nano-biochip applications. Nano Letters 6(4):848–853
Sokuler M, Aronov D et al (2007) Tailored polymer microlenses on treated glass surfaces. Appl Phys Lett 90(20)
Stoll D, Templin MF et al (2002) Protein microarray technology. Front Biosci 7:C13–C32
Taha H, Marks RS et al (2003) Protein printing with an atomic force sensing nanofountain pen. Appl Phys Lett 83(5):1041–1043
Truskett VN, Watts MPC (2006) Trends in imprint lithography for biological applications. Trends Biotechnol 24(7):312–317
Tsarfati-BarAd I, Sauer U et al (2011) Miniaturized protein arrays: model and experiment. Biosens Bioelectron 26(9):3774–3781
Vega RA, Maspoch D et al (2006) Functional antibody arrays through metal ion-affinity templates. Chembiochem 7(11):1653–1657
Vijayendran RA, Leckband DE (2001) A quantitative assessment of heterogeneity for surface-immobilized proteins. Anal Chem 73(3):471–480
Wadu-Mesthrige K, Xu S et al (1999) Fabrication and imaging of nanometer-sized protein patterns. Langmuir 15(25):8580–8583
Wang L, Li PCH (2011) Microfluidic DNA microarray analysis: a review. Anal Chim Acta 687(1):12–27
Xiao PF, He NY et al (2002) In situ synthesis of oligonucleotide arrays by using soft lithography. Nanotechnology 13(6):756–762
Yu X, Xu D et al (2006) Label-free detection methods for protein microarrays. Proteomics 6(20):5493–5503

Part IV
Data Analysis, Commercial and Regulatory Aspects

Chapter 11
Analysis of Array Data and Clinical Validation of Array-Based Assays

Benjamin Haibe-Kains and John Quackenbush

11.1 Introduction

11.1.1 Use of Arrays for Diagnostics and Prediction

In 1995, DNA microarrays were first reported as a tool for probing transcriptional levels on a genomic scale (Lipshutz et al. 1995; Schena et al. 1995), and the research community quickly seized upon this technology as a means of understanding a wide range of biological processes and finding the genes that drive them. Most of the early experiments adopted a simple yet powerful design: comparing two biological classes to identify genes that were differentially expressed between them. Microarrays were used as a tool for gene discovery; many of the applications were to the analysis of gene expression in human cancers (DeRisi et al. 1996; Welford et al. 1998; Khan et al. 1998).

It did not take long for many to realize that arrays could be used for a wide range of applications, including mechanistic studies, finding new subclasses in disease states (Alon et al. 1999; Perou et al. 1999), identifying new biomarkers associated with disease (Moch et al. 1999), and even that the expression patterns themselves could be used as biomarkers to distinguish subclasses of disease (Golub et al. 1999). This latter realization ultimately led to a proliferation of studies that sought patterns of expression that could be used to classify tumor types (Sorlie et al. 2001) and to predict outcome (Beer et al. 2002; van de Vijver et al. 2002) and response to chemotherapy (van't Veer et al. 2002).

B. Haibe-Kains (✉) • J. Quackenbush
Biostatistics and Computational Biology, Dana-Farber Cancer Institute, Boston, MA, USA

Biostatistics, Harvard School of Public Health, Boston, MA, USA
e-mail: bhaibeka@jimmy.harvard.edu; johnq@jimmy.harvard.edu

B. Jordan (ed.), *Microarrays in Diagnostics and Biomarker Development*,
DOI 10.1007/978-3-642-28203-4_11,

Many of the earliest microarray publications on classification of cancer focused on cluster analysis of tumor samples and genes, including applications of hierarchical clustering (Alizadeh et al. 2000; Alon et al. 1999; Bhattacharjee et al. 2001; Perou et al. 1999; Pomeroy et al. 2002; Ross et al. 2000; Sorlie et al. 2001) and partitioning methods such as self-organizing maps (Golub et al. 1999; Pomeroy et al. 2002). These unsupervised data mining approaches have proven useful for class discovery as they take an unbiased approach to finding patterns in the data that can be used to infer the presence of subgroups among the samples. Alizadeh et al. (2000) used hierarchical clustering of cDNA microarray expression data from lymphoma samples to identify two previously unrecognized and transcriptionally distinct subclasses of diffuse large B-cell lymphomas that were eventually related to different stages of B-cell differentiation. One class expressed genes characteristic of germinal B cells (germinal center B-like DLBCL class), while the other expressed genes normally induced during in vitro activation of peripheral blood B cells (activated B-like DLBCL class). Their analysis also showed that patients within these subclasses had distinct clinical prognoses. Perou, Sorlie, and colleagues performed hierarchical clustering of the gene expression profiles of breast tumors and identified 4–5 molecular subtypes (basal-like, HER2-enriched, and luminal-like tumors, the latter which further divided into luminal A, B, and C depending mainly on expression of proliferation genes) (Perou et al. 2000; Sorlie et al. 2003); what has been surprising is the fact that these molecular subtypes recapitulate almost perfectly the current clinical classification of breast tumors based on estrogen receptor (ER) and proliferation (histological grade) and, later, on human epidermal growth factor 2 (HER2) status (Onitilo et al. 2009). Bhattacharjee et al. (2001) used a similar approach to identify lung adenocarcinoma subclasses with different patient outcomes. Ross et al. profiled expression in the 60 cell lines from the National Cancer Institute's anticancer drug screening panel (the NCI 60 cell lines) and used hierarchical clustering to both group cell lines based on their tissue of origin and to find genes that had similar patterns of expression across the samples under the assumption that similar patterns of gene expression could be used to distinguish cell phenotype-specific gene modules (Ross et al. 2000).

It quickly became clear that not only could one use gene expression profiling to explore relationships between existing patient samples, but that one could use gene expression patterns as biomarkers to classify new samples. Golub et al. (1999) were the first to extend the class discovery approach to classification. They used Affymetrix GeneChips™ to profile gene expression on samples from patients suffering from acute lymphoblastic leukemia (ALL) and acute myeloblastic leukemia (AML). The resulting data were analyzed using self-organizing maps (SOMs), and the sample groups they found were highly concordant with the known ALL/AML patient disease groups, suggesting that gene expression could both be used to separate and classify patients. Golub and colleagues then used a weighted gene voting method (a variant on linear discriminant analysis methods) to classify patients' disease using gene expression patterns. They were able to separate patients with a high degree of accuracy, suggesting that expression profiling might provide clinically useful biomarkers for disease classification.

Many other methods have since been applied to using gene expression data to classify patient samples. Ramaswamy et al. (2001) used support vector machines to separate tumors from 14 different organ sites. Bloom et al. (2004) used artificial neural networks to classify 21 different tumor types using data from multiple microarray technologies and further demonstrated that the same approach could be used to predict the primary source for metastatic lesions. Pomeroy et al. (2002) profiled embryonal tumors of the central nervous system (CNS) on Affymetrix GeneChips™ and applied a range of unsupervised and supervised learning methods to investigate whether gene expression data could be used to distinguish between new and existing CNS tumor classes and for patient prognosis. Beer et al. (2002) used an approach based on Cox regression to identify genes for predicting the survival of patients with lung adenocarcinoma.

What these and other studies increasingly demonstrated is that specific patterns of gene expression could be used to discriminate between clinically relevant phenotypes. And from here, it was not difficult to make the leap to using gene expression signatures to predict clinical status or outcome. In many ways, this represents the application with the greatest potential in clinical and translational medicine as it provides new tools for the clinical management of disease. Our goal in this chapter is to describe some of the basic issues related to use of microarray data for disease classification and prognosis, and to highlight some of the potential limitations of this approach.

11.2 Preprocessing

There are many ways to measure gene expression, including any one of a number of array-based technologies, sequence-based approaches, quantitative RT-PCR, and other methods. But fundamentally, each of these, applied to a collection of samples, provides the same basic output—generally a genes-by-samples matrix—that captures an estimate of the transcriptional level of a collection of genes across a set of samples. Before we can use these data, we need to perform basic quality control assays and to compensate for any systematic biases that happen to be introduced in the process, including scaling or "normalizing" the data so that comparisons can be made between samples.

The exact form of such preprocessing of the data depends on the method used to generate the data. As the largest quantity of data is now available from DNA arrays, we will focus our discussion on array-based data. But even here, there are different technologies, including single- and dual-channel (or single- and dual-color) arrays. While the technical details of how one generates the data and performs the preprocessing quality control and normalization vary, the basic principles are universal. Consequently here and throughout the chapter, we will use as an example the analysis of gene expression data collected on the heretofore most widely used platform, the Affymetrix GeneChip™.

11.2.1 Quality Controls

As with all biological assays, assuring the most robust and reliable results from any analysis starts with a focus on the quality and integrity of the underlying data. This includes assuring that the RNA is of good quality and in sufficient quantity, that there is no DNA contamination, that the labeling reactions are efficient and produce sufficient material, and that the hybridizations are carried out in a consistent manner and without significant artifacts such as areas of low intensity on a particular array or obvious failures of a hybridization reaction.

While some of these are easy to avoid experimentally, others require more sophisticated analysis to identify individual arrays of poor quality or entire experimental series that have suboptimal quality. Two types of quality controls for Affymetrix chips are currently widely used:

- Single-chip quality controls: These controls concern one chip at a time. An example is the use of raw image analysis to detect hybridization artifacts like large areas of low intensity due to air bubbles.
- Multichip quality controls: These controls target a set of quantities whose "values should be comparable over all chips of a dataset" (Affymetrix 2004), like scale factors, background intensities, and percentage of present calls. Scale factors are robust measures of the mean level of intensities on a chip. Background intensity is the intensity measured in an empty area (with no hybridization) and returns a measure of the background level. Percentage of present calls measures the proportion of genes being expressed (intensity significantly higher than background) on the chip. Once these quality controls have been carried out, the identification and the consequent removal of the anomalous chips are done.

Note that single-chip controls are well standardized and can be easily performed by the technicians in charge for the hybridizations. This is not the case for multichip quality controls which are more complex and benefitted only recently from the availability of support tools such as R packages, simpleaffy (Wilson and Miller 2005) and affyPLM (Bolstad 2004), and web applications such as AffyGCQC[1] (Osorio et al. 2008) (see McCall et al. (2011a) for a recent review of the existing multichip quality controls). Moreover, large initiatives like the MicroArray Quality Control project (MAQC) provided a series of guidelines regarding quality control and showed that, if performed adequately, it enables better intra- and interplatform reproducibility and has a positive impact on the higher-level analyses, such as clustering or prediction of therapy response (Shi et al. 2006, 2010).

---

[1] http://www.transcriptome.ens.fr/AffyGCQC/.

11.2.2 Normalization

Normalization deals with systematic technical variations between experimental conditions, which are not related to actual biological differences. Normalization methods aim to compensate for these systematic variations between chips in order to enhance the analysis of biological differences between samples. Numerous normalization methods specific to existing gene expression profiling technologies have been proposed in literature. Similarly to quality controls, they can be grouped into two main classes:

- Single-chip normalization methods: These are usually low-complexity methods which use only one chip at a time to define the normalization transformation (e.g., mean scaling). A widely used single-chip normalization for Affymetrix technology is the MicroArray Suite 5 (MAS5) (Affymetrix 2004).
- Multichip normalization methods: These methods use a set of chips to fit a (possibly) complex normalization transformation. This class of methods is sometimes referred to as model-based normalization methods. Widely used multichip normalization methods for Affymetrix technology are the Robust Multichip Average (RMA) (Irizarry et al. 2003a, 2003b), RMA using sequence information (GCRMA) (Wu and Irizarry 2004), DNA-Chip Analyzer (dChip) (Li and Wong 2001), and Variance Stabilization Normalization (VSN) (Huber et al. 2002).

An overview of these normalization methods is given in (Gentleman et al. 2005). Several studies addressed the question of the impact of normalization methods on gene expression analysis (Ploner et al. 2005; Bolstad et al. 2003; Harr and Schlotterer 2006), but no gold standard exists for normalization of Affymetrix data, although RMA is currently used in the majority of the microarray studies.

Two aspects of clinical studies should drive the choice of a normalization method:

- Data sources are usually heterogeneous which could reduce the efficacy of normalization methods. It follows that special care needs to be taken in assessing the quality of a multisource gene expression dataset. In the next subsection, we will discuss in more details which methods can deal with such heterogeneity, also referred to as batch effect.
- The integration of a data-driven predictive tool (such as a gene classifier built from a gene expression dataset) in daily clinical routine is not a trivial task. Since the tool is expected to be used with patients who did not take part in the original study, methods to transform the new patient expression profile into the normalized data space used for building the predictive tool should be taken into account. Until recently, most multichip normalization methods did not provide the possibility to save the fitted model for normalization transformation in order to apply it to new patients. This can be detrimental to the effectiveness of the approach in a real setting where prediction should be computed one

patient at a time, as they come to the clinic. Fortunately, McCall et al. recently developed frozen RMA [fRMA; McCall et al. (2010)], a novel method to efficiently normalize a single microarray experiment at a time while still benefitting from the excellent performance of multichip normalization method that is RMA. The principle behind fRMA is simple but elegant. The authors used the large compendium of Affymetrix datasets available in public databases such as GEO (Barrett et al. 2005) and ArrayExpress (Parkinson et al. 2005) to collect more than 6,000 well-annotated samples. Using this collection of gene expression data, they fitted an updated version of the RMA model for each probeset in the Affymetrix GeneChip and stored these (frozen) normalization models. It is worth to note that by using such a large collection of samples, the authors ensured good generalization of the resulting probeset-specific normalization models. With these models in hand, one can easily use them to normalize a single microarray chip. We will see in the next subsection that fRMA is also beneficial in case of data coming from various sources.

11.2.3 Batch Effects Correction

Batch effects are the systematic nonbiological differences between batches (groups) of samples in microarray experiments due to various causes such as differences in sample preparation and hybridization protocols. While normalization increases the precision of multiarray measurements through the calibration and/or homogenization of the signal intensity distributions, it is not specifically designed for removing such batch effects. Consequently, batch effects may frequently remain after normalization.

The main factors that can cause batch effects (Luo et al. 2010) are:

- The use of different microarray chips (Affymetrix HG-U133A or HG-UPLUS2 for instance) or platforms (Affymetrix, Agilent, or Illumina for instance).
- Even when gene expression profiles are measured using the same microarray chip, the quality of manufacturing may vary from lot to lot.
- Different protocols for managing biological sample; these protocols may vary from center to center and over time within center, which may be a problem for large retrospective studies.
- Different sites or laboratories may have different standard operating procedures (SOPs): RNA isolation, amplification, labeling, and hybridization protocols where different reagents or lots may be used.
- Ambient conditions during sample preparation, such as room temperature and ozone levels.
- Equipment, especially scanner for which types, settings, and calibration may drift over prolonged studies.

Although some of these batch effects could be minimized or even avoided with careful experimental design and appropriate precautions, in many occasions, certain batch effects are unavoidable. For example, many studies require large sample sizes and have to be carried out over many months or years. In other instances, when clinical specimens are involved, experiments are often driven by the availability of the samples which cannot be specifically controlled or accounted for in the original study design, and may originate from a variety of different clinics. Combining data from different batches without carefully removing batch effects can give rise to misleading results, since the bias introduced by the nonbiological nature of the batch effects can be strong enough to mask, or confound true biological differences (Luo et al. 2010).

Normalization methods could be specifically designed to (partially) correct batch effects. This is the case of fRMA, where expression values of probes being strongly affected by the presence of batches will be dramatically reduced (McCall et al. 2010). Moreover, fRMA ensures that each sample has the same distribution of expression values—one of the advantages of the quantile normalization used in RMA and its derivatives (Irizarry et al. 2003b)—which facilitates the joint analysis of data coming from different sources. However, the batch correction implemented in fRMA may not remove all the batch effects, and additional correction techniques should be applied.

Several approaches for batch effect removal have been published in the literature (see (Luo et al. 2010; Chen et al. 2011) for a review and comparison). Commonly used batch effect removal methods include mean/median-centering, standardization, and ratio-based methods where a (set of) chip(s) is used as reference. More sophisticated techniques have been developed during the past 10 years. Among them Alter et al. (2000) applied principal component analysis (PCA) to identify the batch effects and subtract them from the expression measurements. Benito et al. (2004) proposed a method based on distance-weighted discrimination (DWD) which finds a separating hyperplane between two batches and projects the batches onto the DWD plane and subtracts it from the data. Bylesjo et al. (2007) used the orthogonal projections to latent structures method to filter out the latent component that represents the batch effects. Johnson et al. (2007) proposed to use an empirical Bayes approach, called COMBAT,[2] to adjust for batch effects, which pools information across genes and shrinks the batch effect parameter toward the overall mean of the batch estimates across genes.

Discretization techniques can also be used to transform the data in order to remove the batch effects. However, it is difficult to discretize gene expression values in a biologically meaningful way. In real setting, we often do not know which are the natural cutoffs that should be used to discretize gene expressions, and we do not know how many classes are required to well represent the different gene states. Moreover, the discretization should be gene specific which considerably complicates the task. However, Zilliox et al. recently introduced the gene expression barcodes, a method which enables gene-specific binarization of

[2] http://statistics.byu.edu/johnson/ComBat/.

Affymetrix microarray data (Zilliox and Irizarry 2007; McCall et al. 2011b). The idea behind the barcodes is similar to the method used in fRMA. A large collection of well-annotated microarray experiments are collected to accurately estimate the distribution of each gene in various conditions and organisms. A mixture of a Gaussian and a uniform distribution is then fitted to represent the unexpressed and expressed states of each gene, respectively. Using these statistical models on a microarray experiment enables to generate a barcode, a binary representation of the corresponding gene expression profile. Although the resulting barcode is an oversimplification of the microarray data, the authors have shown that the method dramatically attenuates batch effects while yielding good performance for both unsupervised and supervised analysis (Zilliox and Irizarry 2007; McCall et al. 2011b).

11.3 Unsupervised Analysis

Unsupervised analysis enables the discovery of some structure or pattern in the data without supervision, meaning that the analysis is not driven toward prediction of a known phenotype. In this case, a model is built solely on the data in order to predict a class (cancer molecular subtype for instance) without knowing a priori the true classification (Fig. 11.1). Class discovery is widely used in the study of genomic data, notably for identifying new disease subtypes which in turn could be used to better characterize the key biological processes involved in disease genesis and which might yield clinically relevant classifications. However, as we will see in the next section, unsupervised analysis is extremely challenging both in terms of discovery and validation.

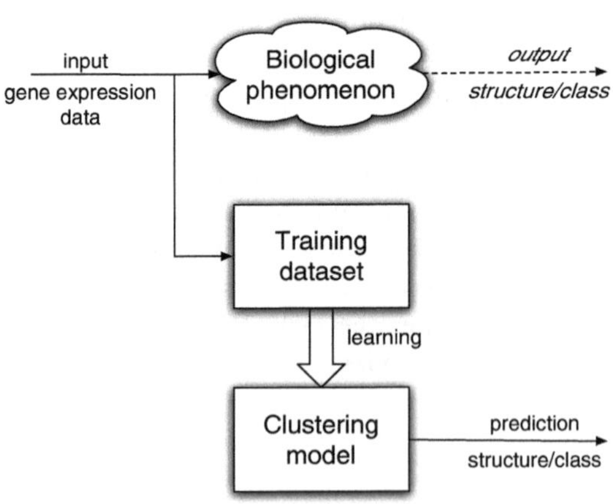

Fig. 11.1 Unsupervised analysis. The clustering model is built using only the genomic array data, without knowing a priori the true classification

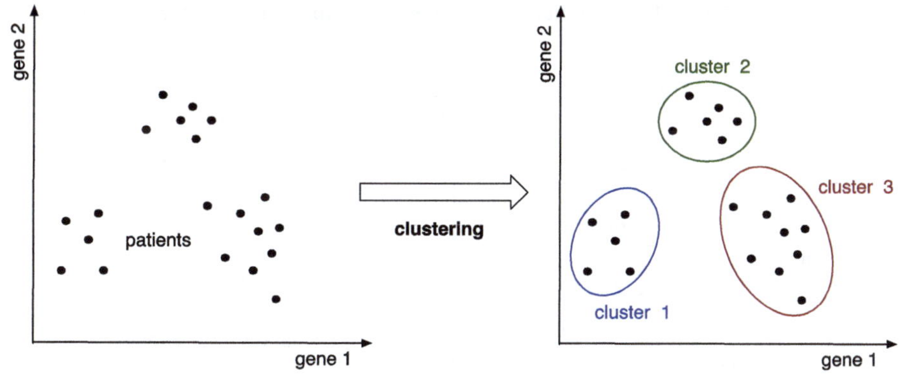

Fig. 11.2 Example of clustering analysis of patient's tumor samples from which expression of two genes has been measured. By performing a clustering analysis, one can identify three clusters, or classes of tumors, sometimes referred to as molecular subtypes; within a cluster, the tumors exhibit similar gene expression profiles (here genes 1 and 2) while being different from tumors in other clusters

11.3.1 Clustering

Class discovery, also referred to as clustering analysis, is the grouping of objects in a population in order to discover some structure in the data. The objects within a group should be similar to one another (those that share some traits such as gene expression) but dissimilar from objects in other groups. Figure 11.2 illustrates such a clustering analysis of tumor samples for which we have measured the expression of two genes.

Clustering is fundamentally a collection of methods of data exploration, often used to assess the presence of natural groupings in the data. However, different methods yield different groupings, since each one implicitly imposes a structure on the data. Moreover, these techniques will produce groupings even if there is no "natural" grouping in the data. Therefore, the analyst must be aware of the structures imposed by the methods and must choose them according to the problem under consideration. One of the main parameters of clustering methods is the choice of a (dis)similarity metrics. As mentioned above, the objects within a cluster should be similar to one another, but dissimilar from objects in other clusters. The definition of the (dis)similarity between objects is based on a notion of distance in the data space such as the Euclidean and correlation-based distances.

There is a vast corpus of literature on clustering and a wide range of application areas. Eisen et al. introduced this methodology in microarray studies (Eisen et al. 1998), but numerous clustering methods have been used since this seminal article: hierarchical clustering (Eisen et al. 1998; de Souto et al. 2008), *k*-means clustering (de Souto et al. 2008), partition around medoids (van der Laan et al. 2003), self-organizing maps (Tamayo et al. 1999), biclustering (Cheng and Church 2000; Sheng et al. 2003), and quality-based clustering (De Smet et al. 2002; Tseng and Wong 2005) to name a few.

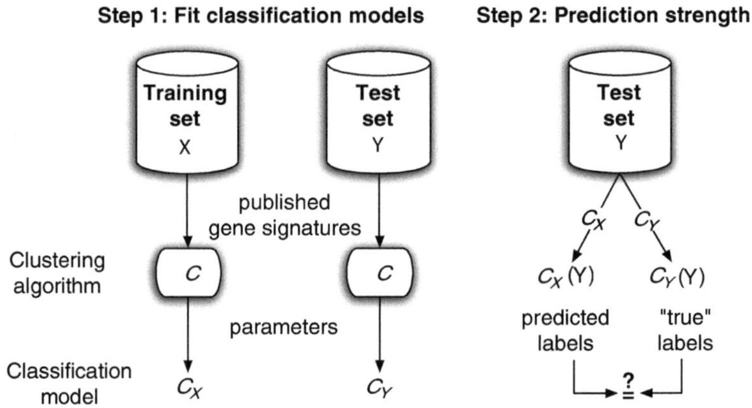

Fig. 11.3 Prediction strength statistic as a measure of performance for class discovery/clustering analysis

11.3.2 Performance Assessment

The performance assessment in clustering analysis is a difficult task since the "truth" remains hidden to the analyst (for unsupervised learning, see Figs. 11.1 and 11.2). However, numerous criteria have been designed in the past few years, most of them allowing for the assessment of the stability/variance of a clustering (Fraley and Raftery 2002; Sugar 1998; Tibshirani et al. 2001; Ben-Hur et al. 2002). These criteria have been mainly used to identify the "good" number of clusters or classes. Recently, Tibshirani and Walther introduced a new framework for performance assessment of clustering analysis (Tibshirani and Walther 2005). The key idea is to view clustering as a supervised classification problem, in which we must also estimate the true class labels. In this framework, the robustness of a clustering model is defined as the capacity of this model to assign the same tumors to the same classes independently of the dataset used to fit it. The resulting prediction strength measure assesses how well the clusters can be predicted from the data.

In Fig. 11.3, we present the idea behind the prediction strength in more details. For each validation cluster defined by the clustering model (referred to as *C*) trained on the test set (*C_Y(Y)*), we compute the proportion of object pairs in that cluster that are also assigned to the same cluster by the clustering model trained on the training set and applied on the test set (*C_X(Y)*). The prediction strength is the minimum of this quantity over the *u* clusters in the validation set and lies in [0, 1]. Based on results obtained from simulated and real data experiments, the authors considered as good a prediction strength $\geq$0.8 (Tibshirani and Walther 2005).

We recently used the prediction strength in a large comparative study of classifiers for breast cancer molecular subtypes of more than 5,000 tumors

(Haibe-Kains et al. 2011). In this study, we showed that a clustering model based on a mixture of Gaussians fitted on only three genes was the most robust (largest prediction strength) and yielded similar prognostic value to more complex classifiers.

11.3.3 Pitfalls and Dangers

Although clustering methods are unavoidable for the visualization and discovery of natural groupings in microarray data, they should be used with caution. Indeed, several issues have emerged since the first publications in the field. First, a clustering method always finds a structure in the data, depending on the choice of the method and the corresponding parameters (such as the number of clusters). Because there is no "truth" or supervision (see Fig. 11.1), it is hard to assess the quality of a clustering and to compare quantitatively different methods, although the prediction strength statistic (Tibshirani and Walther 2005) offers an elegant framework for clustering validation. Second, due to the high feature-to-sample ratio of microarray data, clustering methods may be highly unstable, in that the identification of clusters may strongly depend on the data sample or the gene expressions under consideration. To reduce the risk of overfitting, most analysts attempt to reduce the feature-to-sample ratio by filtering the microarray data in order to remove noisy and uninformative genes. This filtering step, although beneficial, makes the clustering methods more complex to apply.

Additionally, analysts sometimes use clustering methods to perform classification or survival analysis, diverting the original purpose of such methods. In this case, the samples are labeled with respect to the clusters found in the dataset. The use of an unsupervised learning algorithm to perform a supervised task raises important issues, such as the optimization of performance: if a clustering method is used to classify a set of patients, it is not possible to optimize any performance criteria, sensitivity/specificity (Webb 2003), for instance, due to the fact that the clustering method does not use the supervised data (e.g., class or prognosis of the patients) to build a model.

11.4 Supervised Analysis

A supervised analysis consists in finding a relationship—in the form of a statistical model—between input data, such as gene expressions, and output, such as response/resistance to a treatment or the survival of a patient (Fig. 11.4). There are several types of supervised analysis, the choice of which depends on the output of interest: (1) classification analysis if the output is a discrete variable (binary or multiclass), (2) regression analysis if the output is a continuous variable, or

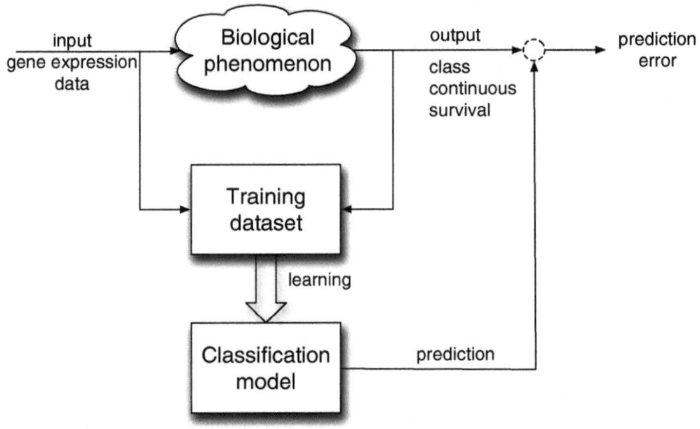

Fig. 11.4 Supervised analysis of gene expression data. The type of statistical model learned from the data, classification, regression, or survival models, depends on the output variable, which can be discrete, continuous, or censored

(3) survival analysis if the output is survival data. These types of analysis will be described in the next sections.

Owing to the complexity and the extremely large dimensionality of genomic array data, a key step in supervised analysis is dimensionality reduction. We describe hereafter the main types of dimensionality reduction techniques and emphasize the properties required for genomic array analysis.

11.4.1 Dimensionality Reduction

Given the high dimensionally of genomic array data, it is not advisable to build a multivariate prediction model using all the variables. Indeed, it is an ill-posed problem since there are too few observations in the dataset to understand the true multivariate relationships between the input and output variables. Therefore, it is beneficial to reduce the dimensionality of the data by either transforming the initial set of input variables (genes) into a reduced set of features (e.g., sets of coexpressed genes often called "metagenes") or to directly select a subset of these variables. Using this reduced set of features, one can subsequently build a multivariate prediction model of lower complexity and consequently less prone to overfitting.

Some properties of feature transformation methods are recommended for genomic array analysis:

- Interpretability: If the final results of the analysis, a gene signature and its corresponding prediction model for instance, have to be interpreted from a biological point of view, the features computed by the feature transformation methods have to be interpretable as well.

- Information: The new features should contain all the "relevant" information from the original input space. The relevance of the new features depends on the biomedical question to address (output; Fig. 11.4). However, it may be difficult to assess the relevance of the information after transformation before completing the whole analysis.
- Generalizability: Most methodologies do not assess the generalizability of feature transformation methods. This may lead to poor performance of the method in an independent dataset since a structure found in one dataset might not be generalizable to another dataset.

We will consider here two types of dimensionality reduction techniques, *feature transformation* and *feature selection*, depending on whether or not the output data is used during the process.

11.4.1.1 Feature Transformation

Feature transformation techniques enable reduction of data dimensionality in an unsupervised way, without using the output data. Three main methods are available for feature transformation: compression, kernel, and clustering methods. Compression and kernel methods transform the original input space into a new one, the dimensions of which are a linear combination of the original variables. These new variables (called features) are difficult to interpret from a biological point of view. Examples of compression and kernel methods are the PCA (Jolliffe and Jolliffe 2002) and the kernel independent component analysis (Bach and Jordan 2003), respectively. Overview of compression and kernel methods is given in (Cristianini et al. 2000; Taylor and Cristianini 2004). Clustering methods rely on the fact that many genes are coexpressed and that their expressions are highly correlated. The approach consists in finding clusters of highly correlated genes and in summarizing each set of clustered genes by the centroid (or prototype) of the cluster (Guyon and Elisseeff 2003). The transformed variables are expected to have lower variance than the original ones, yet remain easy to interpret from a biological and medical point of view.

Feature transformation is expected to be beneficial in terms of better visualization, understanding, and representation of the data. At the same time, it is worth to mention that feature transformation in gene expression datasets exposes the procedure to a risk of overfitting and selection bias if not used in a proper validation framework (Ambroise and McLachlan 2002).

11.4.1.2 Feature Selection

While the role of feature transformation is to reduce the dimensionality of the data in an unsupervised framework, feature selection (Kohavi and John 1997; Guyon and Elisseeff 2003) seeks which variables, among the available ones, provide the largest amount of information about the outcome under study (supervised framework). There are three main categories of feature selection methods: filter, wrapper, and embedded methods. Filter methods assess the relevance of variables ignoring the effects of the selected variable subset on the accuracy of the model. Wrapper methods assess subsets of variables according to their relevance for a given model. The method conducts a search for a "good" subset using the model itself as part of the evaluation function (e.g., forward, backward, and stepwise feature selections). Embedded methods perform feature selection as part of the model fitting and are usually specific to given models (such as classification trees and regularization techniques).

Filter methods, especially variable ranking, are by far the most widely used method in genomic array analysis. Variable ranking assesses the relevance of each individual input variable according to a univariate scoring function supposed to be proportional to the relevance of the variable of interest with respect to the prediction task. Second, it sorts all the variables in a decreasing order according to the value of the scoring function. Eventually, it determines the number of variables to be selected and, subsequently, used in the prediction model, according to some specific criterion. Variable ranking enjoys two interesting properties:

- Computational scalability: It is computationally efficient since it requires only the computation of the n scores, n being the total number of input variables, and the consequent sorting.
- Statistical scalability: Variable ranking, like many filter methods, avoids the estimation of multivariate models to account for the relevance of a set of variables. If on one hand this exposes the technique to some redundancy (large bias), on the other hand, it preserves the approach from overfitting risks (low variance) (Hastie et al. 2001).

However, variable ranking technique requires the definition of a criterion to select the optimal number of features. This is particularly important in a clinical study involving gene expression profiling since this determines the size of the gene signature which is distinctive of the phenotype. At the same time, the task is difficult given the reduced number of samples and the need of using the same dataset for both feature selection and model building. Cross-validation criteria have been proposed in literature (Weiss and Kulikowski 1991) to select the number of variables. Although a cross-validation strategy relies on a multiple fold training and test strategy, it is important to remark that it is still prone to overfitting if it is not kept independent with respect to the model building procedure (Ambroise and McLachlan 2002). For instance, reusing a dataset already used to select a feature set (by cross-validation) in order to assess the quality of a predictive model (again by cross-validation) would return overoptimistic results about the quality of the modeling procedure. A way to minimize the covariance between the selection and the model building procedure would be not to rely on supervised criteria but rather on unsupervised measures. Another limitation of cross-validation criteria is due to the fact that, like other sampling frameworks (bootstrap for instance), it generates different subsets of features for each fold or repetition. This is particularly annoying in a clinical setting where the variability of the selection reduces the confidence of the doctors in the array analysis and casts doubts about the efficacy of the approach.

As an alternative to supervised approaches for signature size selection, recent studies (Pang et al. 2007; Davis et al. 2006) introduced criteria for feature selection stability in genomic array analysis. The idea is to generate multiple datasets using sampling techniques such as cross-validation, random splits, or bootstrapping and to compute for each of these datasets a variable ranking. With all these rankings in hand, the "stability" is estimated by increasing signature sizes (k); such stability criterion is calculated as the proportion of rankings where the top k variables are selected. The "optimal" signature size k depends on the quality of the dataset and the difficulty of the prediction task. In some cases, the top three variables might be highly stable, while in other cases, selection of hundreds of genes is required to maximize the stability of the feature selection.

Although stability reduces only the variance component of the prediction error (expressed conventionally as a bias/variance sum), the large amount of noise and the high dimensionality of the input space suggest that this term could be the most important to address in the bias–variance trade-off. The second advantage deriving from the use of stability measures would be a reinforced confidence of doctors in the gene signature outcomes of clinical studies. This stability-based selection technique has been successfully applied in Davis et al. (2006) and Loi et al. (2008).

11.4.2 Prediction Models

In the following section, we describe classification, regression, and survival analyses in the context of genomic array analysis. Although we will briefly describe the prediction techniques which have been successfully applied to array data, a comprehensive coverage of the underlying mathematical implementation is outside the scope of this chapter; we refer the reader to Hastie et al. (2001), Webb (2003), and Bishop et al. (2006) for further details about the mathematical foundation of prediction models. Given its importance in the field of array analysis, we will describe in more details the existing performance assessment technique for prediction models in the next section.

Independent of the type of analysis (classification regression or survival prediction), the choice of a prediction model depends mainly on its complexity:

- Linear vs nonlinear: Linear models are simpler, more stable than nonlinear models, but not capable of dealing with complex dependencies. On the other hand, the higher complexity of nonlinear models reduces the prediction bias at the cost of an increased variance.
- Univariate vs multivariate: Multivariate models deal more effectively with redundancies than univariate ones but demand ill-conditioned and computationally intensive estimation procedures.

Although complex (multivariate nonlinear) models have been applied in genomic array analysis with the hope of fitting the complex biological nature of diseases, the few array-based models that have been validated in independent studies and have made their way into clinic have low complexity. In breast cancer, for instance, validated prediction models are mostly simple combinations of univariate linear models (Veridex GENE76 (Wang et al. 2005) or the Gene expression Grade Index/MapQuant DX (Sotiriou et al. 2006)) or multivariate linear models with few input variables [Oncotype DX (Paik et al. 2004)]. These results suggest that, when designing a genomic array analysis, one should consider to increase only progressively the complexity of prediction models, using the low-complexity models as benchmark for more complex ones. Indeed, following Occam's razor principle, among the set of models having the best and statistically equivalent performance, the simplest ones should be favored.

In addition, we will see in the section tackling validation of prediction models that simplest models are often easier to translate into datasets composed of different cohorts of patients and generated with different array technologies. Given the scarcity of large microarray datasets with accompanying clinical information, this property is of utmost importance since it facilitates validation of model's performance in independent datasets.

11.4.2.1 Regression

In regression analysis, one is interested in estimating quantitative association between one continuous output variable and a (potentially large) set of input variables. Such output variable may represent a reference biomarker, an indicator of disease, or any variable measuring a clinical response.

There are a great variety of regression techniques for estimating "optimal" mathematical function or model to infer values of a continuous output variable. Artificial neural network (Nepomuceno-Chamorro et al. 2011; Suzuki 2011), regression tree (Breiman et al. 1984; Phuong et al. 2004), and regression support vector machines (Collobert and Bengio 2001) are nonlinear, while Bayesian linear regression (Jin et al. 2008; Sarder et al. 2010) and penalized linear regression model (Tibshirani 2001; Efron et al. 2004; Zou and Hastie 2005) are linear; the latter is the most widely used in genomic array analysis.

Given the high dimensionality of genomic array data, feature transformation and/or feature selection are usually applied prior to fitting of the regression models. The analysts should take into account the computational cost of each regression model; in general, nonlinear models require much more computational resources and time to find "optimal" parameters, and their fitting may be unfeasible for thousands of genes.

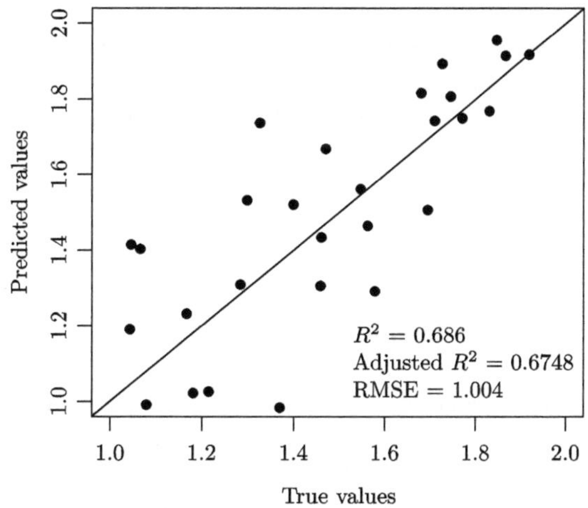

Fig. 11.5 Scatterplot comparing the true values of an output variable with the values predicted by a regression model. In an ideal case, all points would be lined up on the diagonal line. The R^2, adjusted R^2, and the RMSE of the prediction model are given at the bottom right of the plot

Performance Assessment

There exist several criteria for assessing the performance of regression models, but two criteria are well established and considered as "gold standard": (1) the coefficient of determination and (2) the root mean squared error.

The coefficient of determination R^2 is the proportion of variability in a dataset that is accounted for by the prediction model (Steel and Torrie 1980). It provides a measure of how well future outcomes are likely to be predicted by the model. R^2 could be simply calculated as the square of the sample correlation coefficient between the outcomes and their predicted values (Fig. 11.5). Therefore, R^2 ranges from 0 to 1. When analyzing only the training set, one can compute the adjusted R^2 which is a modification of R^2 that adjusts for the number of explanatory terms in a model. Unlike R^2, the adjusted R^2 increases only if the new term improves the model more than would be expected by chance. The adjusted R^2 can be negative and will always be less than or equal to R^2.

A caution that applies to R^2, as to other statistical descriptions of correlation and association, is that "correlation does not imply causation." In other words, while correlations may provide valuable clues regarding causal relationships among variables, a high correlation between two variables does not represent adequate evidence that changing one variable has resulted, or may result, from changes of other variables. Therefore, the fact that a multivariate model fits well the data does not imply that the input variables are the causes of the output variable; for example, genes in the model predicting the survival of cancer patients might not be the cause of tumorigenesis; they could alternatively be effects of cancer progression.

Similar to R^2, the root mean squared error (RMSE) of a regression model is a way to quantify the difference between predicted and the true values of the quantity being estimated (Fig. 11.5). RMSE is always positive and measures the average of the squares of the "errors." The error is the amount by which the predicted value differs from true value. The difference occurs because of randomness or because the prediction model does not account for variables relevant for the prediction of the output variable (Lehmann and Caselia 1998). The RMSE should be minimized; an RMSE of zero, meaning that the regression model predicts observations with perfect accuracy, is the ideal but is practically never possible. Values of RMSE may be used for comparative purposes such that two or more statistical models may be compared using their RMSEs as a measure of how well they explain a given set of observations.

11.4.2.2 Classification

In classification applications, one is interested at predicting a class or category using set of input variables. Typical examples include classification of patients into groups characterized by their response of resistance to a specific targeted therapy, groups based on known tumors' subtypes or mutational status for a gene of interest, etc.

There are plenty of classification models that have been applied to genomic array data. Linear models, such as Fisher's linear discriminant analysis (Fisher 2011), penalized logistic regression (Tibshirani et al. 2001; Zhu and Hastie 2004), naïve Bayes (Berrer et al. 2002; Kelemen et al. 2003), and nearest centroid [prediction analysis of microarrays (Tibshirani et al. 2002)] classifiers, are very popular because of their low complexity, which usually facilitates validation in independent datasets. On the other hand, nonlinear models are also widely used in the field of microarray classification, especially the k-nearest neighbors (Dasarathy 1990; Parry et al. 2010), support vector machines (Furey et al. 2000) and classification trees (Breiman et al. 1984), which include random forests, and their ensemble extension (Manilich et al. 2011; Hu et al. 2008).

As for regression analysis, prior application of dimensionality reduction techniques is highly beneficial to classification performance, whatever the classifier used. Again, simple linear models, such as Fisher's linear discriminant analysis, have been shown to yield performance similar to more complex classifiers in several comparative studies (Dudoit et al. 2002; Lee et al. 2005; Natsoulis et al. 2005).

It is worth to note that the vast majority of the above-mentioned classifiers perform binary classification. For multiclass problems, such as the classification into more than two cancer types or tissue origins, a combination of binary classifiers is usually required. Two techniques exist to make multiclass predictions from binary classifiers: (1) *one*-vs-*all* and (2) *all*-vs-*all*. In one-vs-all, a classifier is built for each class and is trained to discriminate between the class of interest and all the remaining observations; this method is efficient because it requires only the fitting of one classifier per class but it overestimates the variance in the second class by merging potentially very different classes. In all-vs-all, a classifier is built for each pair of classes. It requires to fit many models, but the variance within each class will be estimated properly. We point

Table 11.1 Contingency table for binary classification

		True classes	
		Positive	Negative
Class predictions	Positive	True positive	False positive (type I error)
	Negative	False negative (type II error)	True negative

The table compares the "true" classes with the class predictions

the reader to Rifkin and Klautau (2004) for a review and a comparative study between the two techniques, and to Yeung and Bumgarner (2003) for an application in microarray data analysis.

Performance Assessment

Performance assessment in classifiers is well studied for binary classification where the published performance criteria are mainly based on the notions of sensitivity, specificity, positive and negative predictive values, accuracy, and precision (Table 11.1):

$$\text{Sensitivity} = \frac{\sum \text{True Positive}}{\sum \text{True Positive} + \sum \text{False Negative}},$$

$$\text{Specificity} = \frac{\sum \text{True Negative}}{\sum \text{True Negative} + \sum \text{False Positive}},$$

$$\text{Positive predictive value} = \frac{\sum \text{True Positive}}{\sum \text{True Positive} + \sum \text{False Positive}},$$

$$\text{Negative predictive value} = \frac{\sum \text{True Negative}}{\sum \text{True Negative} + \sum \text{False Negative}},$$

$$\text{Accuracy} = \frac{\sum \text{True Positive} + \sum \text{True Negative}}{\text{Total number of observations}},$$

$$\text{Precision} = \frac{\sum \text{True Positive}}{\sum \text{True Positive} + \sum \text{False Positive}}.$$

If a classifier's output is a continuous value, for example, the probability of class membership as computed by a logistic regression model, the receiving operating characteristic (ROC) curve is a graphical plot of the sensitivity vs (1—specificity) for a classifier as its discrimination threshold varied (Fig. 11.6; Sweets 1988). ROC curve analysis allows analysts to best choose the threshold for the predictive values in order to yield an optimal trade-off between sensitivity and specificity; this can also be done automatically using Youden's index (Youden 1950). A performance

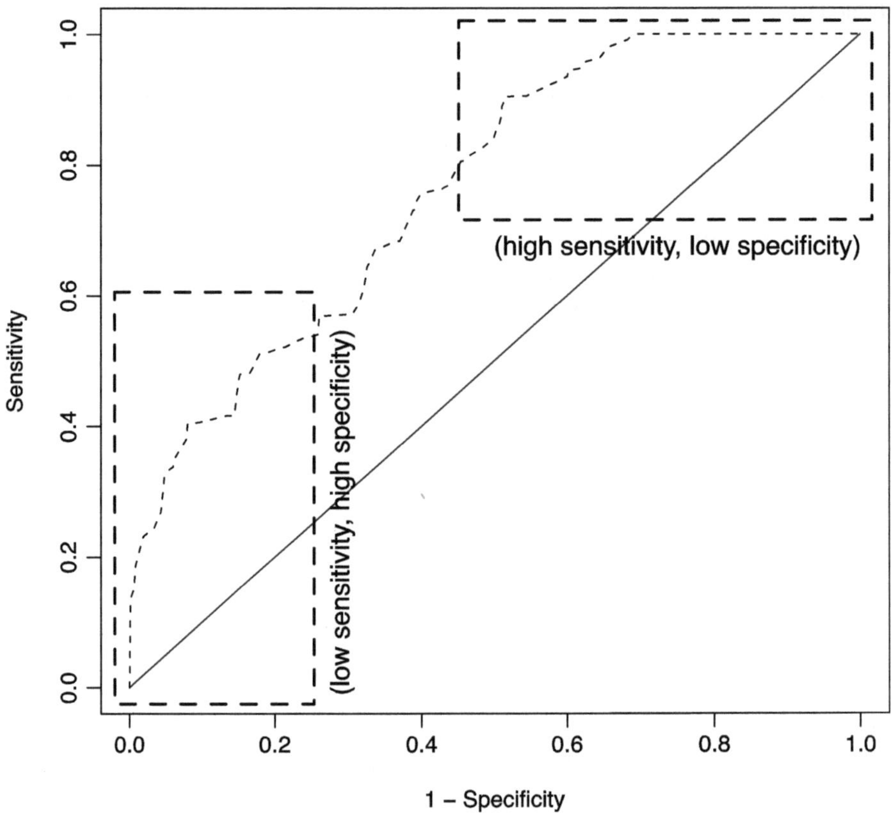

Fig. 11.6 Example of ROC curve. The diagonal line represents the performance of a random model. The curve represents the performance of a prediction score such that large scores stand for prediction of the "higher" (class 1 as opposed to class 0 for instance). The *two boxes* illustrate the regions of the plot where different trade-offs (obtained by applying different cutoffs) can be reached

summary could be obtained by computing the area under the ROC curve [often referred to as AUC or AUROC (Mason and Graham 2002)] which estimates the probability that a classifier will rank a randomly chosen positive instance higher than a randomly chosen negative one. What is particularly interesting about the use of ROC curve to compare different classifiers is that the analyst can identify for which sensitivity/specificity trade-off a classifier outperforms the other(s).

Other performance criteria have been developed in an attempt to deal with class imbalance (much more positive instances than negative ones, or inversely) which may dramatically biased ROC curves, accuracy, or precision. Among them are the squared error, accuracy, and ROC (SAR) score (Caruana and Niculescu-Mizil 2004) and the precision–recall F score (van Rijsbergen 1979).

Performance assessment for multiclass classifiers is more difficult; however, the MAQC workgroup compared different criteria and concluded that Matthews correlation coefficient [MCC (Matthews 1975)] is suitable for microarray analysis in clinical studies (Shi et al. 2010). This criterion is one of the best measures to summarize a contingency table in the presence of class imbalance.

11.4.2.3 Survival Prediction

Unlike classification and regression tasks, survival prediction involves outcome data characterized by event or discontinued observations (Allison and Inc. SI 1995). Let us consider, for instance, 5-year follow-up of a group of cancer patients after the surgical operation. In this case, what is relevant is the dependence between occurrence and timing of the first metastasis and a set of explanatory variables (gene expressions for instance). If we narrow our focus to a binary dependent variable (such as presence or absence of metastasis), conventional classification methods (e.g., logistic regression, linear discriminant analysis, support vector machines) (Duda et al. 2001) would serve our purpose. However, the analysis would ignore the information related to the timing of the event. For instance, it is intuitive to suppose that the aggressiveness of the tumor is related to the time of metastasis appearance during the follow-up (a tumor giving rise to a metastasis during after 6 months is more aggressive than a tumor with a metastasis after 10 years). A possible solution to this problem could come from the adoption of a conventional regression strategy where the time to metastasis plays the role of dependent variable. In this case, the problem would be how to deal with patients for whom no metastasis appeared during the 10 years follow-up. Such cases are referred to as censored. Note that by simply discarding these cases or setting them to a constant value, we would introduce a large bias in our analysis.

Survival analysis combines the information of censored and uncensored data by statistical modeling. The occurring times of events are assumed to be realizations of some random variable T. Two functions are widely used to describe the probability distribution of T:

- The survivor function $S(t) = \Pr\{T > t\}$, where $S(0) = 1$ measuring the probability for an individual to survive until time t.
- The hazard function $\lambda(t) = \lim_{\Delta t \to \infty} \frac{\Pr\left(t \leq T < t + \Delta t | T \geq t\right)}{\Delta_t}$ measuring the instantaneous risk[3] that an event will occur in the interval $[t + \Delta t]$. Note that the probability term is conditional to a survival up to time t.

The survivor function and the hazard function are equivalent ways of describing a continuous probability distribution of survival times (Allison and Inc. SI 1995). The Kaplan–Meier (KM) estimator (also known as the product–limit estimator)

[3] Although it may be helpful to think of the hazard as the instantaneous probability of an event at time t, this quantity is not a probability and may be greater than 1. This is due to the division by Δt. Although the hazard has no upper bound, it cannot be smaller than 0.

returns a nonparametric maximum likelihood estimation of the survivor function (Kaplan and Meier 1958). Semiparametric regression models (Therneau et al. 2000) use the hazard function to model the relationship between the survival outcome and a set of explanatory variables x. A well-known example is the Cox regression model (Cox 1972) which represents the hazard of a patient i as

$$h_i(t) = \lambda_0(t)e^{(\beta_1 x_{i1} + \beta_2 x_{i2} + \ldots + \beta_n x_{in})}.$$

The hazard for individual i at time t, $h_i(t)$, is the product of two factors:

- A baseline hazard function $\lambda_0(t) \geq 0$ which represents the hazard function of an individual where all the explanatory variables are set to zero.
- An exponential function where the argument is a linear function of the explanatory variables. This linear function is also denoted as the linear predictor or the risk score in literature.

Note that while the estimation of the β term is obtained by maximizing a partial likelihood function (Therneau et al. 2000), the estimation of the baseline term is not required. In fact, the fitting of $h_i(t)$ is often done by assuming proportional hazards (the relative risk of event between the two individuals is constant at each interval of time) such that

$$\frac{h_i(t)}{h_j(t)} = e^{(\beta_1(x_{i1} - x_{j1}) + \beta_2(x_{i2} - x_{j2}) + \ldots + \beta_n(x_{in} - x_{jn}))}.$$

Therefore, the hazards between any individual i and j for $i, j \in \{1, 2, \ldots, N\}$ are proportional. An extension of the Cox model allows for multiple strata where patients are classed into disjoint groups, each sharing the estimation of the coefficient β but with a distinct baseline hazard function (Therneau et al. 2000). This is useful in multicenter clinical studies where it is reasonable to assume that different cohorts of patients are characterized by different baseline survival curves.

Building a survival model from expression data is a complex task which put the analyst in front of alternatives similar to regression and classification modeling: linear vs nonlinear and univariate vs multivariate survival models. The nature of gene expression data (very large dimensionality, few samples, and high noise) evokes the potential risks of a nonlinear and multivariate approach. At the same time, a simple univariate model would not be able to account for the multiple interactions underlying the cancer phenomenon.

A plethora of survival models have been developed and applied to genomic array data. Among them, one can cite the artificial neural networks [ANN (Wei et al. 2004)], the support vector machines [Survival-SVM (Van Belle et al. 2011a, b)], the regularized Cox regression model (Tibshirani 1997; Gui and Li 2005; Park and Hastie 2007), random survival forest [RSF (Ulshwaran et al. 2008)], and the semisupervised PCA [SUPERPC (Bair and Tibshirani 2004)]. We refer the reader to (Schumacher et al. 2007; Haibe-Kains et al. 2008; Boulesteix et al. 2008) for more details about the existing methods and how they compare with each other in the context of microarray analysis.

It is worthwhile to note that, although nonlinear multivariate models have been used in the literature, the vast majority of clinical studies relied on simpler approaches and only simple models have been successfully validated in independent studies. In breast cancer, for instance, where the development of gene expression-based prognostic models is an intense research field, the few gene signatures which have been validated in more than one independent dataset are mostly based on linear combination of univariate survival models (Desmedt et al. 2007; van't Veer et al. 2002; Wang et al. 2005; Sotiriou et al. 2006; Finak et al. 2008; Loi et al. 2005, 2010; Desmedt et al. 2008). The interest of such additive modeling schemes (Hastie and Tibshirani 1990) lies in the fact that the linear combination of several univariate models returns a model which is simple, yet able to address multivariate tasks.

Performance Assessment

In contrast to classification and regression frameworks where performance criteria are well established, performance assessment in survival analysis is not trivial because it lacks standards. Therefore, diverse performance criteria have been developed, while none of them reached a consensus, each research group having its favorite method to assess the quality of a survival model.

However, theoretical studies attempted to characterize the properties of a good performance criterion for survival analysis [see (Royston and Sauerbrei 2004) for instance], while comparative studies sought to assess the impact of the use of different performance criteria for model selection and comparison (Schumacher et al. 2007; Haibe-Kains et al. 2008; Boulesteix et al. 2008).

Royston and Sauerbrei defined a set of properties that a "good" performance criterion should possess (Royston and Sauerbrei 2004):

- Interpretability: The performance criterion should have a simple and intuitively appealing meaning.
- Generality: The performance criterion should be applicable to risk score and risk group predictions.
- Directedness: When the risk ordering changes, the performance criterion should change in the appropriate direction. For example, if the risk ordering were reversed, then the performance criterion should change sign.
- Unbiasedness: The observed value of the performance criterion should be an unbiased estimate of the true value. In particular, the expectation of the performance criterion should be close to a known value when the proposed risk ordering bears on average no relationship with the true risk ordering. This may occur when the model is useless, or less commonly, when the dataset holds no information relevant for the prediction task.

- Responsiveness: If the risk gradient were reduced, for example, by omission of a covariate highly relevant for the prediction task, then the performance criterion should move substantially toward a known value.
- Robustness: The performance criterion should not be unduly influenced by a small number of extreme risk predictions (outliers).
- Precision: Confidence intervals for the performance criterion should be computed straightforwardly.
- Independence of censoring: The performance criterion should be independent of the degree of censoring in the data.

We will describe here the existing performance criteria for survival analysis and briefly compare them to the hazard ratio which is by far the most widely used in clinical studies using microarray data to predict patients' prognosis or benefit from therapies.

The hazard ratio can be defined as a summary of the difference between two survival curves, representing the reduction in the risk of event by unit or between two different groups. It is a form of relative risk. The hazard ratio possesses the properties of generality, directedness, responsiveness, and precision. However, it suffers from some drawbacks:

- Since the hazard ratio is estimated through Cox's regression, the proportional hazards assumption must hold. The impact of the departure from this assumption is difficult to assess, although it does not seem to affect the hazard ratio estimation dramatically (Cox 1972).
- The interpretation of the hazard ratio depends on the scale of the input variable (continuous or discrete), since this ratio represents the difference in risk per unit and per time interval. For instance, let r be a risk lying in $[-1, 1]$; a hazard ratio of 2 means that the hazard of a patient of risk $r = 1$ is twice the hazard of a patient of risk $r = 0$ and four times the hazard of a patient of risk $r = -1$. The scale of the input variable is not always interpretable, rendering difficult the interpretation of the hazard ratio.
- The estimation of the hazard ratio is not robust to the presence of outliers.

These drawbacks raise the question of the use of alternative performance criteria in survival analysis.

The D index is a measure of separation between the hazard function of each patient (Royston and Sauerbrei 2004). It is a robust estimate of the log hazard ratio comparing two equal-sized risk groups. The main advantage of the D index compared to the hazard ratio is that the transformation of the input values in ranks allows for comparing performance of different datasets without calibration since the risk scores have the same scales and its interpretation does not depend on the scale of the input variable anymore. Moreover, the D index is robust to outliers. D index possesses all the properties of a "good" performance criterion.

The cross-validated partial likelihood has originally been introduced to optimize the regularization parameter in the penalized Cox regression model (Verweij and van Houwelingen 1993; van Houwelingen et al. 2006), the cross-validated partial likelihood estimates, in a cross-validation framework, the partial likelihood of a Cox model fitted with the predictions as input variable. This performance criterion has the advantage of using the same quantity as the cost function in Cox's regression, that is, the partial likelihood. However, there is no easy interpretation of this performance criterion, and it is not robust to the presence of outliers. The cross-validated partial likelihood possesses the properties of generality and responsiveness.

The Brier score, denoted by BSC, is defined as the squared difference between an event occurrence (0/1) and its predicted probabilities at time t. Probabilities of event can be derived from Cox's model fitted with the predictions as input variable. The intuition behind the Brier score is that, if a patient experiences no event at time t, the predicted probability of event occurrence should be close to 0; symmetrically if the patient experiences an event, the probability should be close to 1. This performance criterion has a form very similar to the mean squared error, widely used in regression (Hastie et al. 2001). The main drawback of the Brier score is that it requires to estimate the baseline hazard function in case of Cox's model, which is a difficult task (Therneau et al. 2000). Moreover, this performance criterion is not robust to the presence of outliers. The Brier score possesses the properties of generality and responsiveness.

The concordance index computes the probability that, for a pair of randomly chosen comparable patients, the patient with the higher risk prediction will experience an event before the lower risk patient (or inversely). It is a generalization of the area under ROC curve (AUC) to deal with the presence of censoring on the survival data (Harrell et al. 1996). In contrast to the hazard ratio and D index, the concordance makes no assumption about the hazard of the patients. It is robust to outliers since only the order of the risk predictions matters. Its interpretation is simple and equivalent to the area under a ROC curve, well known in supervised classification theory (Hastie et al. 2001). Moreover, a recent modification of the traditional concordance index formulation makes this performance criterion insensitive to the censoring distribution of the survival data (Uno et al. 2011).

Time-dependent ROC curves: The receiver operating characteristic (ROC) curve is a standard technique for assessing the performance of a continuous variable for binary classification (Sweets 1988). As a ROC curve is a plot of sensitivity versus $1 -$ specificity for all the possible cutoff values of a continuous predictive score, Heagerty et al. used the Kaplan–Meier estimator (Kaplan and Meier 1958) into the definition of sensitivity and specificity in order to take into account censoring present in survival data (Heagerty et al. 2000). This performance criterion has the same advantages as the concordance index, including having no assumptions and being robust to outliers, and is particularly useful for interpretation. Indeed, the analyst is able to visually assess the trade-off between sensitivity and specificity for the classifications computed by applying each possible cutoff (Fig. 11.6). The AUC of a time-dependent ROC curve possesses the properties of interpretability, directedness, unbiasedness, responsiveness, and robustness.

11.4.3 Pitfalls and Dangers

In contrast to unsupervised analysis, the "truth" (outcome values, either continuous, class labels, or survival) is known in supervised analysis such that it is used to build and optimize prediction models. This enables the development of complex prediction rules, which can reasonably fit the biology underlying the biomedical problem of interest. However, one must be extremely cautious when fitting a prediction model as this process has been shown to be prone to overfitting (Ambroise and McLachlan 2002; Simon 2003; Michiels et al. 2005; Gamberger and Lavrac 2004; Dupuy and Simon 2007). This is due to the intrinsic characteristics of array data: (1) high feature-to-sample ratio (typically few hundreds observations for several thousands of genes), (2) high level of noise in the measurements, and (3) redundancy and co-(non)linearity present in the data. To reduce risk of overfitting, the complexity of the prediction models should be increased only progressively since simpler models are often more generalizable and could yield similar or even outperform more complex prediction rules (see Haibe-Kains et al. (2008) for breast cancer prognosis). Moreover, a proper validation framework should be implemented to robustly assess the performance of a prediction model in independent datasets (Ambroise and McLachlan 2002; Michiels et al. 2005; Dupuy and Simon 2007).

11.5 Clinical Validation

The translation of an array-based diagnostic or predictive tool from the initial discovery in the lab to clinic (*from bench to bedside*) is a long, difficult path but, if successful, leads to sustainable solutions for public health. A recent study from Contopoulos-Ioannidis and colleagues reported that, from the first description of the discovery to licensed clinical use by 2003, the median period for translating a (set of) biomarker(s) into clinic is 24 years (interquartile range of 14–44 years), also referred to as "translational lag" (Contopoulos-Ioannidis et al. 2008). It is worth to note that the authors focused on the very highly cited papers (>1,000 citations) to increase the likelihood of seeing a discovery validated and translated into clinical routine. The translation lag was even longer for those studies which were not fully reproduced by independent studies (17 years vs 44 years for fully and not fully reproducible studies, respectively). These results suggest that, although the translation process is lengthy, it could be significantly accelerated by ensuring full reproducibility of the initial discovery and by facilitating validation on independent datasets, also referred to as portability.

The pipeline of clinical assay development is given in Fig. 11.7. The first three steps are usually part of the initial publication reporting the biomarkers, the technology used to measure them, and their promising performance, which suggest the relevance of their translation into clinics (steps 1–3). An assay is then designed

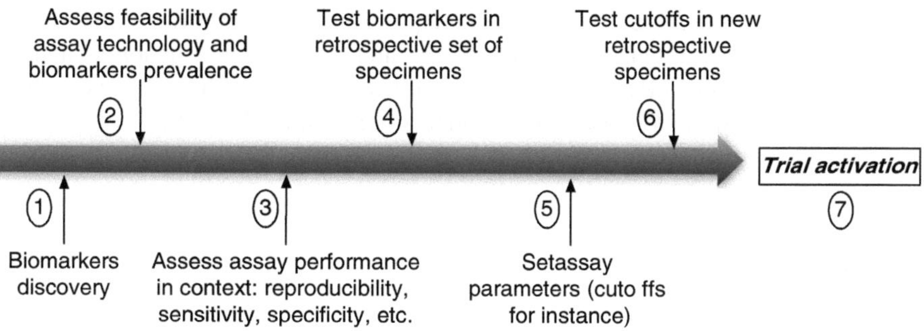

Fig. 11.7 Clinical assay development pipeline, figure inspired from http://cadp.cancer.gov/about_cadp/

and validated in a retrospective set of specimens to characterize the analytical and clinical validity (steps 4–6). Finally, the clinical utility of the assay is determined in a clinical trial, the last step of the assay development pipeline (step 7).

11.5.1 Reproducibility

Genomic array data analysis is a complex process, involving many steps and methodological choices. As a consequence, these analyses are difficult to reproduce and prone to errors if not properly documented. Full reproducibility of array analysis is becoming more and more important; the retraction of several high-impact publications from Dr. Anil Potti, former professor at Duke University, is a recent example illustrating how the lack of reproducibility may be surrogate for artifactual discoveries and may prevent translation of any assay into clinics (The Cancer Letter 2011).

The reproduction of wet lab experiments is a difficult process, but it can be greatly facilitated with comprehensive description of the experimental protocols and materials. Full replication of dry lab experiments, which are genomic array data analyses in our case, is possible though; there exist guidelines and tools to help researchers make their analyses documented and fully reproducible (Mesirov 2010; Gentleman 2005). Sweave is a well-established tool in the bioinformatics research community to ensure reproducibility of complex analyses; it relies on a combination of R code[4] and LaTeX typesetting[5] to automatically generate, from the raw genomic array data, the results and figures reported in the corresponding

[4] http://www.r-project.org/.

[5] http://www.latex-project.org/.

publication (Leisch 2002). Tools which require less programming skills are currently under development (Mesirov 2010), and widely used genomic analysis software, such as GenePattern,[6] now integrates modules which enable exportation of data and all parameters used during analysis (Reich et al. 2006).

11.5.2 Portability

The transition from the initial discovery to the implementation of an efficient assay is a lengthy, challenging task. To facilitate portability of the initial findings, we must ensure that (1) the discovered biomarkers are robust and their performance validated in independent datasets and (2) that the prediction model using those biomarkers could be applied using a technology which is optimized in terms of cost and reliability for clinical use.

The validation of biomarkers in multiple independent datasets often requires a prediction model developed with a specific array platform (Affymetrix GeneChip for instance) to be applied on another platform (in-house microarray chip, Agilent and Illumina for instance). This is far from trivial, and this should be taken into account during design of the analytical pipeline. Indeed, initial dataset could be simplified—summarization of several probes representing the same gene into one value to avoid annotation ambiguities—in order to facilitate validation in different array platforms. Moreover, models, such as combination of univariate prediction rules, are easier to translate from one array platform to another and are less sensitive to differences in data preprocessing. In contrast, complex prediction models may not be so successful: a multivariate Cox model could not be applied if one variable (gene) is not present in another platform and its performance could be dramatically reduced if the (linear or nonlinear) nature of interaction between genes is different.

The design of a clinical assay often requires choosing a technology that is different from the one used during the initial development of the prediction model. This is due to the fact that one can focus on a reduced set of biomarkers in contrast to the discovery phase; this opens new perspective in terms of costs and reliability. A typical example would be the translation of a gene expression microarray into less complex qRT-PCR assay.

11.5.3 Analytical Validity

In developing clinical assays, a critical element is establishing the ability of the assay to accurately and reproducibly measure the quantity of the relevant analytes. Before the introduction of array-based technologies, most clinical assays measured

[6] http://www.broadinstitute.org/cancer/software/genepattern/.

a small number of proteins, and the important measure for those proteins—and the one used in a clinical setting—was their concentrations in blood, urine, or other bodily fluids.

For single-analyte assays, establishing analytical validity is relatively straightforward. Repeated measures on a standard dilution series can provide the necessary data to establish both the quantitative nature of an assay and its reproducibility. Such data can actually be used algorithmically to correct the measurements in an assay so that one can accurately estimate analyte concentration.

However, how one establishes analytical validity for compound biomarkers involving genomic assays remains an unsolved problem. The challenge in array-based nucleic acid hybridization is that the assay itself is widely recognized as being semiquantitative at best. Although efforts such as the MicroArray Quality Control project (MAQC) have addressed the development of new standards for establishing the analytical validity of microarray assays (Shi et al. 2006, 2010), these have not yet been widely adopted.

Despite these efforts, it remains somewhat unclear whether the ability to accurately measure the absolute quantity of one or more analytes is essential for microarray-based biomarkers. Most microarray biomarkers rely not on absolute measures of gene expression levels, but on either the relative expression of various genes compared to each other or on the overall pattern of expression of multiple genes. The global pattern of expression provides multiple sources of internal control and, as applied, the absolute expression of individual genes rarely if ever considered. While analytical validity may ultimately be important as these tests become standardized, their clinical validity may ultimately more important.

11.5.4 Clinical Validity

The ultimate test of a biomarker is its ability to accurately and reproducibly predict clinically relevant outcome—and this is true of array-based biomarkers. Although we have discussed the value of using independent datasets to validate a biomarker, this is only the first step to performing a rigorous clinical validation. Although the approaches for clinical validation of genomic biomarkers are still being developed, the “gold standard” approach would be to assess their predictive ability in a randomized clinical trial.

The details of such a trial would depend on the nature of the biomarker being evaluated. The challenge is that genomic biomarkers are increasingly being developed based on clinical endpoints such as response to particular therapies. In such a case, the temptation would be to screen individual patients and assign them to treatment or not based on their individual genomic profiles. In such a design, the natural endpoint would be the overall response of the patients as measured by some clinical parameter such as survival. However, such a design and the interpretation of the results are potentially confounded by the convolution of treatment and assignment.

A far cleaner and more justifiable approach would be to conduct a trial in two stages. In the first stage, one could randomly assign patients to treatment or nontreatment arms for the drug of interest independent of the patients' genomic profile. This would be a way or evaluating the efficacy of the treatment independent of other variables. The genomic profiling data could then be used to evaluate whether one could predict response in the treatment arm and to do a sensitivity/specificity analysis. Assuming that this was encouraging and that there were benefits to predicted responders who received the treatment, one could then move to a second stage trial in which patients were assigned to treatment arms based on the genomic profiles.

While such studies are difficult and time consuming, they do provide an opportunity to establish the clinical validity and clinical utility of genomic biomarkers. Unfortunately, very few genomic biomarkers have made the transition from the laboratory to the clinic, but those that have either cleared the hurdle of clinical validation or are being considered in large-scale clinical trials to establish their clinical validity are promising (Slodkowska and Ross 2009; Straver et al. 2010; Ross et al. 2008; Cardoso et al. 2007; Mook et al. 2007; Mamounas et al. 2008; Marchionni et al. 2007; Eng-Wong and Zujewski 2008; Habel et al. 2006).

11.5.5 Clinical Utility

While the validity of a biomarker is important, one other critical aspect that must be considered is its utility. The critical question is whether the biomarker can be used effectively to improve the treatment and management of patients in a meaningful way that can improve outcome or quality of life.

For example, if a biomarker can partition patients into good and poor prognosis groups, but there is no reasonable therapy or change in treatment protocol that can be used with the good or poor prognosis groups, then the biomarker in question is not likely to be widely adopted as a clinical assay because it adds cost to treatment without any potential benefit. One can consider the example of MammaPrint (van't Veer et al. 2002) and Oncotype DX (Cobleigh et al. 2005), both genomic signature biomarkers for predicting recurrence in node-negative breast cancer. While neither of these are used at directing therapies, both can be used to help patients and their physicians to decide whether to pursue adjuvant chemotherapy or instead to adopt a course of disease surveillance.

The truth is that genomic biomarkers are still in their early days and the community is struggling to establish guidelines to assist in determining whether they are appropriate for clinical use (Kohli-Laven et al. 2011; Kelley et al. 2011). However, what is clear is that these assays are going to continue to be developed and that many of them will be making their way into clinical use over the coming years.

11.6 Conclusions

The launch of the Human Genome Project in the 1990s and its announced completion in 2000 generated great excitement about the potential for improving our understanding and treatment of disease. The explosion of gene expression studies using arrays provided the first real opportunity to see this vision realized, but only a limited number of studies have produced actionable outcomes with relevance to clinical medicine.

There are many reasons for this, but probably the most significant problem is that we did not recognize the complexity of the problem we were tackling. Many studies assumed that disease states were nearly homogenous, that the causes were tied to common genetic factors, and that strong biological signals would emerge from genomic analyses of relatively small numbers of samples. Increasingly, we are coming to understand that diseases often have multiple subtypes defined by their genetic backgrounds and genomic profiles—and that these subtypes must be treated independently to avoid confounding their analysis.

We also recognize the need for extensive clinical annotation of the samples we analyze so we can avoid confounding effects and ground our analyses in clinically relevant endpoints. The field as a whole has also begun to appreciate the need for sound experimental design and for good statistical and computational analysis, as well as understanding that there is a need for new statistical and analytical methods that account for the complex relationships between genomic elements and which can place results into a biological framework.

But the most important and underappreciated problem has been a shortage of high-quality, well-annotated data. Despite the investment of millions of dollars and the generation of thousands of genomic assays, the number which can be effectively reused to develop and validate genomic biomarkers remains surprisingly small. Nevertheless, the genomic biomarkers that have made their way into clinical application provide a roadmap for future development.

And it is likely that the necessary data will be available in the future. Although nearly all of the data that has been generated is array based, the cost of new sequencing technologies has been falling nearly 33% per quarter since 2007. At this pace, whole-genome sequencing, exome sequencing, RNA-seq, and other genomic profiling techniques are likely to become more widespread. And as the data become less expensive to generate and more widely available, the growing number of patients for whom we have genomic data will open new opportunities for development of robust and reliable genomic biomarkers—provided we apply the lessons we have learned from our experience with array-based methods.

References

Affymetrix (2004) *GeneChip expression analysis: data analysis fundamentals*, vol 2447, pp 1–42. doi:10.1002/jnr.10268

Alizadeh AA, Eisen MB, Davis RE, Ma C, Lossos IS, Rosenwald A, Boldrick JC, Sabet H, Tran T, Yu X, Powell JI, Yang L, Marti GE, Moore T, Hudson J Jr, Lu L, Lewis DB, Tibshirani R, Sherlock G, Chan WC, Greiner TC, Weisenburger DD, Armitage JO, Warnke R, Levy R, Wilson W, Grever MR, Byrd JC, Botstein D, Brown PO, Staudt LM (2000) Distinct types of diffuse large B-cell lymphoma identified by gene expression profiling. Nature 403(6769): 503–511

Allison PD, Inc. SI (eds) (1995) Survival analysis using SAS: a practical guide. SAS Institute Inc., Cary, NC

Alon U, Barkai N, Notterman DA, Gish K, Ybarra S, Mack D, Levine AJ (1999) Broad patterns of gene expression revealed by clustering analysis of tumor and normal colon tissues probed by oligonucleotide arrays. Proc Natl Acad Sci USA 96(12):6745–6750

Alter O, Brown PO, Botstein D (2000) Singular value decomposition for genome-wide expression data processing and modeling. Proc Natl Acad Sci USA 97(18):10101–10106. doi:97/18/10101 [pii]

Ambroise C, McLachlan GJ (2002) Selection bias in gene extraction on the basis of microarray gene-expression data. Proc Natl Acad Sci USA 99(10):6562–6566. doi:10.1073/pnas.102102699

Bach FR, Jordan MI (2003) Kernel independent component analysis. J Mach Learn Res 3:1–48

Bair E, Tibshirani R (2004) Semi-supervised methods to predict patient survival from gene expression data. PLoS Biol 2(4):511–522

Barrett T, Suzek TO, Troup DB, Wilhite SE, Ngau WC, Ledoux P, Rudnev D, Lash AE, Fujibuchi W, Edgar R (2005) NCBI GEO: mining millions of expression profiles—database and tool. Nucleic Acids Res 33:D562

Beer DG, Kardia SL, Huang CC, Giordano TJ, Levin AM, Misek DE, Lin L, Chen G, Gharib TG, Thomas DG, Lizyness ML, Kuick R, Hayasaka S, Taylor JM, Iannettoni MD, Orringer MB, Hanash S (2002) Gene-expression profiles predict survival of patients with lung adenocarcinoma. Nat Med 8(8):816–824

Ben-Hur A, Elisseeff A, Guyon I (2002) A stability based method for discovering structure in clustered data. Proc Pac Symp Biocomput 7:6–17

Benito M, Parker J, Du Q, Wu J, Xiang D, Perou CM, Marron JS (2004) Adjustment of systematic microarray data biases. Bioinformatics 20(1):105–114

Berrer DP, Dubitzky W, Granzow M (2002) A practical approach to microarray data analysis, 1st edn. Springer, New York

Bhattacharjee A, Richards WG, Staunton J, Li C, Monti S, Vasa P, Ladd C, Beheshti J, Bueno R, Gillette M, Loda M, Weber G, Mark EJ, Lander ES, Wong W, Johnson BE, Golub TR, Sugarbaker DJ, Meyerson M (2001) Classification of human lung carcinomas by mRNA expression profiling reveals distinct adenocarcinoma subclasses. Proc Natl Acad Sci USA 98(24):13790–13795

Bishop CM, Jordan M, Kleinberg J, Scholkopf B (eds) (2006) Pattern recognition and machine learning information science and statistics. Springer, New York

Bloom G, Yang IV, Boulware D, Kwong KY, Coppola D, Eschrich S, Quackenbush J, Yeatman TJ (2004) Multi-platform, multi-site, microarray-based human tumor classification. Am J Pathol 164(1):9–16

Bolstad BM (2004) Low-level analysis of high-density oligonucleotide array data: background normalization and summarization. University of California, Berkeley

Bolstad BM, Irizarry RA, Astrand M, Speed TP (2003) A comparison of normalization methods for high density oligonucleotide array data based on variance and bias. Bioinformatics 19(2):185–193

Boulesteix AL, Porzelius C, Daumer M (2008) Microarray-based classification and clinical predictors: on combined classifiers and additional predictive value. Bioinformatics 24(15):1698–1706. doi:btn262 [pii]10.1093/bioinformatics/btn262

Breiman L, Friedman JH, Olshen RA, Stone CJ (1984) Classification and regression trees. Chapman and Hall, New York

Bylesjo M, Eriksson D, Sjodin A, Jansson S, Moritz T, Trygg J (2007) Orthogonal projections to latent structures as a strategy for microarray data normalization. BMC Bioinformatics 8:207. doi:1471-2105-8-207 [pii]10.1186/1471-2105-8-207

Cardoso F, Piccart-Gebhart M, Van't Veer L, Rutgers E (2007) The MINDACT trial: the first prospective clinical validation of a genomic tool. Mol Oncol 1(3):246–251. doi:S1574-7891(07)00077-4 [pii]10.1016/j.molonc.2007.10.004

Caruana R, Niculescu-Mizil A (2004) Data mining in metric space: an empirical analysis of supervised learning performance criteria. Paper presented at the ACM SIGKDD international conference on Knowledge discovery and data mining, New York

Chen C, Grennan K, Badner J, Zhang D, Gershon E, Jin L, Liu C (2011) Removing batch effects in analysis of expression microarray data: an evaluation of six batch adjustment methods. PLoS One 6(2):e17238. doi:10.1371/journal.pone.0017238

Cheng Y, Church GM (2000) Biclustering of expression data. Proc Int Conf Intell Syst Mol Biol 8:93–103

Cobleigh MA, Tabesh B, Bitterman P, Baker J, Cronin M, Liu ML, Borchik R, Mosquera JM, Walker MG, Shak S (2005) Tumor gene expression and prognosis in breast cancer patients with 10 or more positive lymph nodes. Clin Cancer Res 11(24 Pt 1):8623–8631. doi:11/24/8623 [pii]10.1158/1078-0432.CCR-05-0735

Collobert R, Bengio S (2001) SVMTorch: support vector machines for large-scale regression problems. J Mach Learn Res 1:143–160

Contopoulos-Ioannidis DG, Alexiou GA, Gouvias TC, Ioannidis JP (2008) Medicine. Life cycle of translational research for medical interventions. Science 321(5894):1298–1299. doi:321/5894/1298 [pii]10.1126/science.1160622

Cox DR (1972) Regression models and life tables. J R Stat Soc Ser B 34:187–220

Cristianini N, Press CCU, Shawe-Taylor J (eds) (2000) An introduction to support vector machines and other kernel-based learning methods. Cambridge University Press, Cambridge

Dasarathy BV (ed) (1990) Nearest neighbor: pattern classification techniques. IEEE Computer Society Press, New York

Davis CA, Gerick F, Hintermair V, Friedel CC, Fundel K, Kuffner R, Zimmer R (2006) Reliable gene signatures for microarray classification: assessment of stability and performance. Bioinformatics 22(19):2356–2363. doi:10.1093/bioinformatics/btl400

De Smet F, Mathys J, Marchal K, Thijs G, De Moor B, Moreau Y (2002) Adaptive quality-based clustering of gene expression profiles. Bioinformatics 18(5):735–746

de Souto M, Costa I, de Araujo D, Ludermir T, Schliep A (2008) Clustering cancer gene expression data: a comparative study. BMC Bioinformatics 9(1):497. doi:10.1186/1471-2105-9-497

DeRisi J, Penland L, Brown PO, Bittner ML, Meltzer PS, Ray M, Chen Y, Su YA, Trent JM (1996) Use of a cDNA microarray to analyse gene expression patterns in human cancer. Nat Genet 14(4):457–460

Desmedt C, Piette F, Loi SM, Wang Y, Lallemand F, Haibe-Kains B, Viale G, Delorenzi M, Zhang Y, d'Assignies MS, Bergh J, Lidereau R, Ellis P, Harris AL, Klijn JGM, Foekens JA, Cardoso F, Piccart MJ, Buyse M, Sotiriou C, Consortium T (2007) Strong time dependence of the 76-gene prognostic signature for node-negative breast cancer patients in the TRANSBIG multicenter independent validation series. Clin Cancer Res 13(11):3207–3214

Desmedt C, Haibe-Kains B, Wirapati P, Buyse M, Larsimont D, Bontempi G, Delorenzi M, Piccart M, Sotiriou C (2008) Biological processes associated with breast cancer clinical outcome depend on the molecular subtypes. Clin Cancer Res 14(16):5158–5165. doi:10.1158/1078-0432.CCR-07-4756

Duda RO, Hart PR, Stork DG (2001) Pattern classification. Wiley, New York

Dudoit S, Fridlyand J, Speed TP (2002) Comparison of discrimination methods for the classification of tumors using gene expression data. J Am Stat Assoc 97(457):77–87

Dupuy A, Simon RM (2007) Critical review of published microarray studies for cancer outcome and guidelines on statistical analysis and reporting. J Natl Cancer Inst 99(2):147–157. doi:10.1093/jnci/djk018

Efron B, Hastie T, Johnstone I, Tibshirani R (2004) Least angle regression. Ann Stat 32(2):407–499

Eisen M, Spellman P, Brown P, Botstein D (1998) Cluster analysis and display of genome-wide expression patterns. PNAS 95:14863–14868

Eng-Wong J, Zujewski JA (2008) Current NCI-sponsored cooperative group trials of endocrine therapies in breast cancer. Cancer 112(3 Suppl):723–729. doi:10.1002/cncr.23188

Finak G, Bertos N, Pepin F, Sadekova S, Souleimanova M, Zhao H, Chen H, Omeroglu G, Meterissian S, Omeroglu A, Hallett M, Park M (2008) Stromal gene expression predicts clinical outcome in breast cancer. Nat Med 14(5):518–527. doi:10.1038/nm1764

Fisher RA (2011) The use of multiple measurements in taxonomic problems. Ann Eugen 7:179–188

Fraley C, Raftery AE (2002) Model-based clustering, discriminant analysis, and density estimation. J Am Stat Assoc 97(458):611–631. doi:10.1198/016214502760047131

Furey TS, Cristianini N, Duffy N, Bednarski DW, Schummer M, Haussler D (2000) Support vector machine classification and validation of cancer tissue samples using microarray expression data. Bioinformatics 16(10):906–914

Gamberger D, Lavrac N (2004) Avoiding data overfitting in scientific discovery: experiments in functional genomics. Paper presented at the ECAI, 22–27 Aug 2004, Valencia, Spain

Gentleman R (2005) Reproducible research: a bioinformatics case study. Stat Appl Genet Mol Biol 4(1)

Gentleman R, Huber W, Carey VJ, Irizarry RA, Dudoit S (2005) Bioinformatics and computational biology solutions using R and bioconductor. Springer, New York

Golub TR, Slonim DK, Tamayo P, Huard C, Gaasenbeek M, Mesirov JP, Coller H, Loh ML, Downing JR, Caligiuri MA, Bloomfield CD, Lander ES (1999) Molecular classification of cancer: class discovery and class prediction by gene expression monitoring. Science 286 (5439):531–537

Gui J, Li H (2005) Penalized Cox regression analysis in the high-dimensional and low-sample size settings, with applications to microarray gene expression data. Bioinformatics 21(13):3001–3008. doi:10.1093/bioinformatics/bti422

Guyon I, Elisseeff A (2003) An introduction to variable and feature selection. J Mach Learn Res 3:1157–1182

Habel LA, Shak S, Jacobs MK, Capra A, Alexander C, Pho M, Baker J, Walker M, Watson D, Hackett J, Blick NT, Greenberg D, Fehrenbacher L, Langholz B, Quesenberry CP (2006) A population-based study of tumor gene expression and risk of breast cancer death among lymph node-negative patients. Breast Cancer Res 8(3):R25. doi:bcr1412 [pii]10.1186/bcr1412

Haibe-Kains B, Desmedt C, Sotiriou C, Bontempi G (2008) A comparative study of survival models for breast cancer prognostication based on microarray data: does a single gene beat them all? Bioinformatics 24(19):2200–2208. doi:10.1093/bioinformatics/btn374

Haibe-Kains B, Desmedt C, Loi SM, Culhane AC, Bontempi G, Quackenbush J, Sotiriou C (2012) A three-gene model to robustly identify breast cancer molecular subtypes. J Natl Cancer Inst 104(4):311–325. doi:10.1093/jnci/djr545

Harr B, Schlotterer C (2006) Comparison of algorithms for the analysis of Affymetrix microarray data as evaluated by co-expression of genes in known operons. Nucleic Acids Res 34(2):8

Harrell FJ, Lee K, Mark D (1996) Multivariable prognostic models: issues in developing models, evaluating assumptions and adequacy, and measuring and reducing errors. Stat Med 15(4):361–387. doi:10.1002/(SICI)1097-0258(19960229)15:4<361::AID-SIM168>3.0.CO;2-4

Hastie T, Tibshirani R (1990) Generalized additive models. Chapman and Hall, London

Hastie T, Bickel P, Tibshirani R, Diggle P, Friedman J, Fienberg S, Gather U, Otkin I, Zeger S (eds) (2001) The elements of statistical learning statistics. Springer, New York

Heagerty PJ, Lumley T, Pepe MS (2000) Time-dependent ROC curves for censored survival data and a diagnostic marker. Biometrics 56:337–344

Hu H, Li J-Y, Wang H, Daggard G, Wang L-Z (2008) Robustness analysis of diversified ensemble decision tree algorithms for Microarray data classification. Paper presented at the 2008 International Conference on Machine Learning and Cybernetics (ICMLC), Kunming, 12–15 Jul 2008

Huber W, von Heydebreck A, Sultman H, Poustka A, Vingron M (2002) Variance stabilization applied to microarray data calibration and to the quantification of differential expression. Bioinformatics 18(1):S96–S104

Irizarry RA, Boldstad BM, Collin F, Cope LM, Hobbs B, Speed TR (2003a) Summaries of affymetrix GeneChip probe level data. Nucleic Acids Res 31(4)

Irizarry RA, Hobbs B, Collin F, Beazer-Barclay YD, Antonellis KJ, Scherf U, Speed TP (2003b) Exploration, normalization, and summaries of high density oligonucleotide array probe level data. Biostatistics (Oxford, England) 4(2):249–264. doi:10.1093/biostatistics/4.2.249

Jin R, Si L, Chan C (2008) A Bayesian framework for knowledge driven regression model in micro-array data analysis. Int J Data Min Bioinform 2(3):250–267

Johnson WE, Li C, Rabinovic A (2007) Adjusting batch effects in microarray expression data using empirical Bayes methods. Biostatistics 8(1):118–127. doi:10.1093/biostatistics/kxj037

Jolliffe IT, Jolliffe IT (eds) (2002) Principal component analysis. Springer series in statistics. Springer, New York

Kaplan EL, Meier P (1958) Nonparametric estimation from incomplete observations. J Am Stat Assoc 53:451–457

Kelemen A, Zhou H, Lawhead P, Liang Y (2003) Naive Bayesian classifier for microarray data. In: 2003 International joint conference on neural networks, vol 3, pp 1769–1773. Paper presented at the 2003 international joint conference on neural networks, IEEE. doi:10.1109/IJCNN.2003.1223675

Kelley RK, Wang G, Venook AP (2011) Biomarker use in colorectal cancer therapy. J Natl Compr Canc Netw 9(11):1293–1302. doi:9/11/1293 [pii]

Khan J, Simon R, Bittner M, Chen Y, Leighton SB, Pohida T, Smith PD, Jiang Y, Gooden GC, Trent JM, Meltzer PS (1998) Gene expression profiling of alveolar rhabdomyosarcoma with cDNA microarrays. Cancer Res 58(22):5009–5013

Kohavi R, John GH (1997) Wrappers for feature subset selection. Artif Intell 97(1–2):273–324

Kohli-Laven N, Bourret P, Keating P, Cambrosio A (2011) Cancer clinical trials in the era of genomic signatures: biomedical innovation, clinical utility, and regulatory-scientific hybrids. Soc Stud Sci 41(4):487–513

Lee JW, Lee JB, Park M, Song SH (2005) An extensive comparison of recent classification tools applied to microarray data. Computational Statistics & Data Analysis 48(4):869–885. doi: 10.1016/j.csda.2004.03.017

Lehmann EL, Caselia G (1998) Theory of point estimation, 2nd edn. Springer, New York

Leisch F (2002) Sweave. Dynamic generation of statistical reports using literate data analysis. In: Computational statistics, vol 69, pp 575–580. Presented at the computational statistics, SFB adaptive information systems and modelling in economics and management science, WU Vienna University of Economics and Business. http://www.google.ca/url?sa=t&rct=j&q=sweave.%20dynamic%20generation%20of%20statistical%20reports%20using%20literate%20data%20analysis&source=web&cd=1&ved=0CDQQFjAA&url=http%3A%2F%2Fepub.wu.ac.at%2F1788%2F1%2Fdocument.pdf&ei=qiVNT7TPLevTiALGwp2wDw&usg=AFQjCNGZ5hg-vOqrB2j6hU7HGhQkhiBrRg&sig2=dmMu57Xag5ci-fANUqxnAA

Li C, Wong WH (2001) Model-based analysis of oligonucleotide arrays: model validation, design issues and standard error application. Genome Biol 2(8):1–11

Lipshutz RJ, Morris D, Chee M, Hubbell E, Kozal MJ, Shah N, Shen N, Yang R, Fodor SP (1995) Using oligonucleotide probe arrays to access genetic diversity. Biotechniques 19(3):442–447

Loi SM, Haibe-Kains B, Desmedt C, Wirapati P, Lallemand F, Tutt AM, Gillet C et al (2008) Predicting prognosis using molecular profiling in estrogen receptor-positive breast cancer treated with tamoxifen. BMC Genomics 9:239. doi:10.1186/1471-2164-9-239

Loi SM, Haibe-Kains B, Desmedt C, Wirapati P, Lallemand F, Tutt AM, Gillet C, Ellis P, Ryder K, Reid JF, Daidone MG, Pierotti MA, Berns EM, Jansen MP, Foekens JA, Delorenzi M,

Bontempi G, Piccart MJ, Sotiriou C (2008) Predicting prognosis using molecular profiling in estrogen receptor-positive breast cancer treated with tamoxifen. BMC Genomics 9:239. doi:10.1186/1471-2164-9-239

Loi SM, Haibe-Kains B, Majjaj S, Lallemand F, Durbecq V, Larsimont D, Gonzalez-Angulo AM, Pusztai L, Symmans WF, Bardelli A, Ellis P, Tutt ANJ, Gillett CE, Hennessy BT, Mills GB, Phillips WA, Piccart MJ, Speed TP, McArthur GA, Sotiriou C (2010) PIK3CA mutations associated with gene signature of low mTORC1 signaling and better outcomes in estrogen receptor-positive breast cancer. Proc Natl Acad Sci USA 107(22):10208–10213. doi:10.1073/pnas.0907011107

Luo J, Schumacher M, Scherer A, Sanoudou D, Megherbi D, Davison T, Shi T, Tong W, Shi L, Hong H, Zhao C, Elloumi F, Shi W, Thomas R, Lin S, Tillinghast G, Liu G, Zhou Y, Herman D, Li Y, Deng Y, Fang H, Bushel P, Woods M, Zhang J (2010) A comparison of batch effect removal methods for enhancement of prediction performance using MAQC-II microarray gene expression data. Pharmacogenomics J 10(4):278–291. doi:tpj201057 [pii]10.1038/tpj.2010.57

Mamounas E, Budd GT, Miller K (2008) Incorporating the oncotype DX breast cancer assay into community practice: an expert Q and A and case study sampling. Clin Adv Hematol Oncol 6 (2):s1–s8

Manilich EA, Ozsoyoglu ZM, Trubachev V, Radivoyevitch T (2011) Classification of large microarray datasets using fast random forest construction. J Bioinform Comput Biol 9 (2):251–267. doi:S021972001100546X [pii]

Marchionni L, Wilson RF, Marinopoulos SS, Wolff AC, Parmigiani G, Bass EB, Goodman SN (2007) Impact of gene expression profiling tests on breast cancer outcomes. Evid Rep Technol Assess (Full Rep) 160:1–105

Mason SJ, Graham NE (2002) Areas beneath the relative operating characteristics (ROC) and relative operating levels (ROL) curves: statistical significance and interpretation. Q J R Meteorol Soc 128(584):2145–2166. doi:10.1256/003590002320603584

Matthews BW (1975) Comparison of the predicted and observed secondary structure of T4 phage lysozyme. Biochim Biophys Acta 405(2):442–451

McCall MN, Bolstad BM, Irizarry RA (2010) Frozen robust multiarray analysis (fRMA). Biostatistics (Oxford, England) 11(2):242–253. doi:10.1093/biostatistics/kxp059

McCall MN, Murakami PN, Lukk M, Huber W, Irizarry RA (2011a) Assessing affymetrix GeneChip microarray quality. BMC Bioinformatics 12:137. doi:1471-2105-12-137 [pii] 10.1186/1471-2105-12-137

McCall MN, Uppal K, Jaffee HA, Zilliox MJ, Irizarry RA (2011b) The gene expression barcode: leveraging public data repositories to begin cataloging the human and murine transcriptomes. Nucleic Acids Res 39(Database issue):D1011–D1015. doi:gkq1259 [pii]

Mesirov JP (2010) Computer science accessible reproducible research. Science 327(5964): 415–416. doi:327/5964/415 [pii]10.1126/science.1179653

Michiels S, Koscielny S, Hill C (2005) Prediction of cancer outcome with microarrays: a multiple random validation strategy. Lancet 365:488–492

Moch H, Schraml P, Bubendorf L, Mirlacher M, Kononen J, Gasser T, Mihatsch MJ, Kallioniemi OP, Sauter G (1999) Identification of prognostic parameters for renal cell carcinoma by cDNA arrays and cell chips. Verh Dtsch Ges Pathol 83:225–232

Mook S, van't Veer LJ, Rutgers EJ, Piccart-Gebhart MJ, Cardoso F (2007) Individualization of therapy using mammaprint: from development to the MINDACT Trial. Cancer Genomics Proteomics 4(3):147–155

Natsoulis G, El Ghaoui L, Lanckriet GRG, Tolley AM, Leroy F, Dunlea S, Eynon BP, Pearson CI, Tugendreich S, Jarnagin K (2005) Classification of a large microarray data set: algorithm comparison and analysis of drug signatures. Genome Res 15(5):724–736. doi:10.1101/gr.2807605

Nepomuceno-Chamorro I, Azuaje F, Devaux Y, Nazarov PV, Muller A, Aguilar-Ruiz JS, Wagner DR (2011) Prognostic transcriptional association networks: a new supervised approach based on regression trees. Bioinformatics 27(2):252–258. doi:btq645 [pii]10.1093/bioinformatics/btq645

Onitilo AA, Engel JM, Greenlee RT, Mukesh BN (2009) Breast cancer subtypes based on ER/PR and Her2 expression: comparison of clinicopathologic features and survival. Clin Med Res 7(1–2):4–13. doi:10.3121/cmr.2009.825

Osorio YFJ, Prina E, Lang T, Milon G, Davory C, Coppee JY, Regnault B (2008) AffyGCQC: a web-based interface to detect outlying genechips with extreme studentized deviate tests. J Bioinform Comput Biol 6(2):317–334. doi:S0219720008003400 [pii]

Paik S, Shak S, Tang G, Kim C, Bakker J, Cronin M, Baehner FL, Walker MG, Watson D, Park T, Hiller W, Fisher ER, Wickerham DL, Bryant J, Wolmark N (2004) A multigene assay to predict recurrence of tamoxifen-treated, node-negative breast cancer. N Engl J Med 351(27):2817–2826

Pang S, Havukkala I, Hu Y, Kasabov N (2007) Classification consistency analysis for bootstrapping gene selection. Neural Comput Appl 18(6):527–539

Park MY, Hastie T (2007) L1 regularization path algorithm for generalized linear models. J R Stat Soc 69:659–677

Parkinson H, Sarkans U, Shojatalab M, Contrino S, Coulson R, Farne A, Lara GG, Holloway E, Kapushesky M, Lilja P, Mukherjee G, Oezcimen A, Rayner T, Rocca-Serra P, Sharma A, Sansone S, Brazma A (2005) ArrayExpress: a public repository for microarray gene expression data at the EBI. Nucleic Acids Res 33:D553–D555

Parry RM, Jones W, Stokes TH, Phan JH, Moffitt RA, Fang H, Shi L, Oberthuer A, Fischer M, Tong W, Wang MD (2010) k-Nearest neighbor models for microarray gene expression analysis and clinical outcome prediction. Pharmacogenomics J 10(4):292–309. doi:10.1038/tpj.2010.56

Perou CM, Jeffrey SS, van de Rijn M, Rees CA, Eisen MB, Ross DT, Pergamenschikov A, Williams CF, Zhu SX, Lee JC, Lashkari D, Shalon D, Brown PO, Botstein D (1999) Distinctive gene expression patterns in human mammary epithelial cells and breast cancers. Proc Natl Acad Sci USA 96(16):9212–9217

Perou CM, Sorlie T, Eisen MB, van de Rijn M, Jeffrey SS, Rees CA, Pollack JR, Ross DT, Johnsen H, Akslen LA, Fluge O, Pergamenschikov A, Williams C, Zhu SX, Lonning PE, Borresen-Dale A-L, Brown PO, Botstein D (2000) Molecular portraits of human breast tumours. Nature 406(6797):747–752. doi:10.1038/35021093

Phuong TM, Lee D, Lee KH (2004) Regression trees for regulatory element identification. Bioinformatics 20(5):750–757. doi:10.1093/bioinformatics/btg480 btg480 [pii]

Ploner A, Miller LD, Hall P, Bergh J, Pawitan Y (2005) Correlation test to assess low-level processing of high-density oligonucletide microarray data. BMC Bioinformatics 6(80):1–20

Pomeroy SL, Tamayo P, Gaasenbeek M, Sturla LM, Angelo M, McLaughlin ME, Kim JY, Goumnerova LC, Black PM, Lau C, Allen JC, Zagzag D, Olson JM, Curran T, Wetmore C, Biegel JA, Poggio T, Mukherjee S, Rifkin R, Califano A, Stolovitzky G, Louis DN, Mesirov JP, Lander ES, Golub TR (2002) Prediction of central nervous system embryonal tumour outcome based on gene expression. Nature 415(6870):436–442

Ramaswamy S, Tamayo P, Rifkin R, Mukherjee S, Yeang CH, Angelo M, Ladd C, Reich M, Latulippe E, Mesirov JP, Poggio T, Gerald W, Loda M, Lander ES, Golub TR (2001) Multiclass cancer diagnosis using tumor gene expression signatures. Proc Natl Acad Sci USA 98(26):15149–15154

Reich M, Liefeld T, Gould J, Lerner J, Tamayo P, Mesirov JP (2006) GenePattern 2.0. Nat Genet 38(5):500–501. doi:10.1038/ng0506-500

Rifkin R, Klautau A (2004) In defense of One-Vs-All classification. J Mach Learn Res 5(1):101–141

Ross DT, Scherf U, Eisen MB, Perou CM, Rees C, Spellman P, Iyer V, Jeffrey SS, Van de Rijn M, Waltham M, Pergamenschikov A, Lee JC, Lashkari D, Shalon D, Myers TG, Weinstein JN, Botstein D, Brown PO (2000) Systematic variation in gene expression patterns in human cancer cell lines. Nat Genet 24(3):227–235

Ross JS, Hatzis C, Symmans WF, Pusztai L, Hortobagyi GN (2008) Commercialized multigene predictors of clinical outcome for breast cancer. Oncologist 13(5):477–493. doi:13/5/477 [pii] 10.1634/theoncologist.2007-0248

Royston P, Sauerbrei W (2004) A new measure of prognostic separation in survival data. Stat Med 23(5):723–748. doi:10.1002/sim.1621

Sarder P, Schierding W, Cobb JP, Nehorai A (2010) Estimating sparse gene regulatory networks using a bayesian linear regression. IEEE Trans Nanobioscience 9(2):121–131. doi:10.1109/TNB.2010.2043444

Schena M, Shalon D, Davis RW, Brown PO (1995) Quantitative monitoring of gene expression patterns with a complementary DNA microarray. Science 270(5235):467–470

Schumacher M, Binder H, Gerds TA (2007) Assessment of survival prediction models based on microarray data. Bioinformatics 23(14):1768–1774

Sheng Q, Moreau Y, De Moor B (2003) Biclustering microarray data by Gibbs sampling. Bioinformatics 19(Suppl 2):ii196–ii205. doi:10.1093/bioinformatics/btg1078

Shi L, Reid LH, Jones WD, Shippy R, Warrington JA, Baker SC, Collins PJ, de Longueville F, Kawasaki ES, Lee KY, Luo Y, Sun YA, Willey JC, Setterquist RA, Fischer GM, Tong W, Dragan YP, Dix DJ, Frueh FW, Goodsaid FM, Herman D, Jensen RV, Johnson CD, Lobenhofer EK, Puri RK, Schrf U, Thierry-Mieg J, Wang C, Wilson M, Wolber PK, Zhang L, Amur S, Bao W, Barbacioru CC, Lucas AB, Bertholet V, Boysen C, Bromley B, Brown D, Brunner A, Canales R, Cao XM, Cebula TA, Chen JJ, Cheng J, Chu TM, Chudin E, Corson J, Corton JC, Croner LJ, Davies C, Davison TS, Delenstarr G, Deng X, Dorris D, Eklund AC, Fan XH, Fang H, Fulmer-Smentek S, Fuscoe JC, Gallagher K, Ge W, Guo L, Guo X, Hager J, Haje PK, Han J, Han T, Harbottle HC, Harris SC, Hatchwell E, Hauser CA, Hester S, Hong H, Hurban P, Jackson SA, Ji H, Knight CR, Kuo WP, LeClerc JE, Levy S, Li QZ, Liu C, Liu Y, Lombardi MJ, Ma Y, Magnuson SR, Maqsodi B, McDaniel T, Mei N, Myklebost O, Ning B, Novoradovskaya N, Orr MS, Osborn TW, Papallo A, Patterson TA, Perkins RG, Peters EH, Peterson R, Philips KL, Pine PS, Pusztai L, Qian F, Ren H, Rosen M, Rosenzweig BA, Samaha RR, Schena M, Schroth GP, Shchegrova S, Smith DD, Staedtler F, Su Z, Sun H, Szallasi Z, Tezak Z, Thierry-Mieg D, Thompson KL, Tikhonova I, Turpaz Y, Vallanat B, Van C, Walker SJ, Wang SJ, Wang Y, Wolfinger R, Wong A, Wu J, Xiao C, Xie Q, Xu J, Yang W, Zhang L, Zhong S, Zong Y, Slikker W Jr (2006) The MicroArray quality control (MAQC) project shows inter- and intraplatform reproducibility of gene expression measurements. Nat Biotechnol 24(9):1151–1161

Shi L, Campbell G, Jones WD, Campagne F, Wen Z, Walker SJ, Su Z, Chu TM, Goodsaid FM, Pusztai L, Shaughnessy JD Jr, Oberthuer A, Thomas RS, Paules RS, Fielden M, Barlogie B, Chen W, Du P, Fischer M, Furlanello C, Gallas BD, Ge X, Megherbi DB, Symmans WF, Wang MD, Zhang J, Bitter H, Brors B, Bushel PR, Bylesjo M, Chen M, Cheng J, Chou J, Davison TS, Delorenzi M, Deng Y, Devanarayan V, Dix DJ, Dopazo J, Dorff KC, Chou J, Davison TS, Delorenzi M, Deng Y, Devanarayan V, Dix DJ, Dopazo J, Dorff KC, Elloumi F, Fan J, Fan S, Fan X, Fang H, Gonzaludo N, Hess KR, Hong H, Huan J, Irizarry RA, Judson R, Juraeva D, Lababidi S, Lambert CG, Li L, Li Y, Li Z, Lin SM, Liu G, Lobenhofer EK, Luo J, Luo W, McCall MN, Nikolsky Y, Pennello GA, Perkins RG, Philip R, Popovici V, Price ND, Qian F, Scherer A, Shi T, Shi W, Sung J, Thierry-Mieg D, Thierry-Mieg J, Thodima V, Trygg J, Vishnuvajjala L, Wang SJ, Wu J, Wu Y, Xie Q, Yousef WA, Zhang L, Zhang X, Zhong S, Zhou Y, Zhu S, Arasappan D, Bao W, Lucas AB, Berthold F, Brennan RJ, Buness A, Catalano JG, Chang C, Chen R, Cheng Y, Cui J, Czika W, Demichelis F, Deng X, Dosymbekov D, Eils R, Feng Y, Fostel J, Fulmer-Smentek S, Fuscoe JC, Gatto L, Ge W, Goldstein DR, Guo L, Halbert DN, Han J, Harris SC, Hatzis C, Herman D, Huang J, Jensen RV, Jiang R, Johnson CD, Jurman G, Kahlert Y, Khuder SA, Kohl M, Li J, Li M, Li QZ, Li S, Liu J, Liu Y, Liu Z, Meng L, Madera M, Martinez-Murillo F, Medina I, Meehan J, Miclaus K, Moffitt RA, Montaner D, Mukherjee P, Mulligan GJ, Neville P, Nikolskaya T, Ning B, Page GP, Parker J, Parry RM, Peng X, Peterson RL, Phan JH, Quanz B, Ren Y, Riccadonna S, Roter AH, Samuelson FW, Schumacher MM, Shambaugh JD, Shi Q, Shippy R, Si S, Smalter A, Sotiriou C, Soukup M, Staedtler F, Steiner G, Stokes TH, Sun Q, Tan PY, Tang R, Tezak Z, Thorn B, Tsyganova M, Turpaz Y, Vega SC, Visintainer R, von Frese J, Wang C, Wang E, Wang J, Wang W, Westermann F, Willey JC, Woods M, Wu S, Xiao N, Xu J, Xu L, Yang L, Zeng X, Zhang M, Zhao C, Puri RK, Scherf U, Tong W, Wolfinger RD, Consortium M (2010) The MicroArray quality control (MAQC)-II study of common practices for the development and validation of microarray-based predictive models. Nat Biotechnol 28(8):827–838. doi:nbt.1665 [pii]

Simon R (2003) Diagnostic and prognostic prediction using gene expression profiles in high-dimensional microarray data. Br J Cancer 89:1599–1604

Slodkowska EA, Ross JS (2009) MammaPrint 70-gene signature: another milestone in personalized medical care for breast cancer patients. Expert Rev Mol Diagn 9(5):417–422. doi:10.1586/erm.09.32

Sorlie T, Perou CM, Tibshirani R, Aas T, Geisler S, Johnsen H, Hastie T, Eisen MB, van de Rijn M, Jeffrey SS, Thorsen T, Quist H, Matese JC, Brown PO, Botstein D, Eystein Lonning P, Borresen-Dale AL (2001) Gene expression patterns of breast carcinomas distinguish tumor subclasses with clinical implications. Proc Natl Acad Sci USA 98(19):10869–10874

Sorlie T, Tibshirani R, Parker J, Hastie T, Marron JS, Nobel A, Deng S, Johnsen H, Pesich R, Geister S, Demeter J, Perou C, Lonning PE, Brown PO, Borresen-Dale A-L, Botstein D (2003) Repeated observation of breast tumor subtypes in independent gene expression data sets. Proc Natl Acad Sci USA 1(14):8418–8423

Sotiriou C, Wirapati P, Loi S, Harris A, Fox S, Smeds J, Nordgren H, Farmer P, Praz V, Haibe-Kains B, Desmedt C, Larsimont D, Cardoso F, Peterse H, Nuyten D, Buyse M, Van de Vijver MJ, Bergh J, Piccart M, Delorenzi M (2006) Gene expression profiling in breast cancer: understanding the molecular basis of histologic grade to improve prognosis. J Natl Cancer Inst 98(4):262–272

Steel RGD, Torrie JH (1980) Principles and procedures of statistics. McGraw Hill, New York

Straver ME, Glas AM, Hannemann J, Wesseling J, van de Vijver MJ, Rutgers EJ, Vrancken Peeters MJ, van Tinteren H, van't Veer LJ, Rodenhuis S (2010) The 70-gene signature as a response predictor for neoadjuvant chemotherapy in breast cancer. Breast Cancer Res Treat 119(3):551–558. doi:10.1007/s10549-009-0333-1

Sugar C (1998) Techniques for clustering and classification with applications to medical problems. Doctoral Thesis, Stanford University

Suzuki K (ed) (2011) Artificial neural networks—methodological advances and biomedical applications. Artifical Neural Network Intech, Croatia

Sweets JA (1988) Measuring the accuracy of diagnostic systems. Science 240(4857):1285–1293

Tamayo P, Slonim D, Mesirov J, Zhu Q, Kitareewan S, Dmitrovsky E, Lander ES, Golub TR (1999) Interpreting patterns of gene expression with self-organizing maps: methods and application to hematopoietic differentiation. Proc Natl Acad Sci USA 96(6):2907–2912

Taylor JS, Cristianini N (2004) Kernel methods for pattern analysis. Cambridge University Press, New York

The Cancer Letter (2011) Duke accepts potti resignation; retraction process initiated with nature medicine. http://www.cancerletter.com/articles/20101123_1

Therneau TM, Gail M, Grambsch PM, Krickeberg K, Samet JM, Tsiatis A, Wong W (eds) (2000) Modeling survival data: extending the Cox model. Statistics for biology and health. Springer, New York. doi:10.1002/sim.956

Tibshirani R (1997) The lasso method for variable selection in the Cox model. Stat Med 16(4): 385–395. doi:10.1002/(SICI)1097-0258(19970228)16:4<385::AID-SIM380>3.0.CO;2-3 [pii]

Tibshirani R (2001) Regression shrinkage and selection via the lasso. J Royal Statist Soc B 58(1):1267–1288

Tibshirani R, Walther G (2005) Cluster validation by prediction strength. J Comput Graph Stat 14(3):511–528

Tibshirani R, Walther G, Hastie T (2001) Estimating the number of clusters in a data set via the gap statistic. J R Stat Soc Ser B Stat Methodol 63(2):411–423. doi:10.1111/1467-9868.00293

Tibshirani R, Hastie T, Narasimhan B, Chu G (2002) Diagnosis of multiple cancer types by shrunken centroids of gene expression. Proc Natl Acad Sci USA 99(10):6567–6572. doi:10.1073/pnas.082099299 99/10/6567 [pii]

Tseng GC, Wong WH (2005) Tight clustering: a resampling-based approach for identifying stable and tight patterns in data. Biometrics 61(1):10–16. doi:10.1111/j.0006-341X.2005.031032.x

Ulshwaran H, Kogalur U, Blackstone E, Lauer M (2008) Random survival forests. Ann Appl Stat 2(3):841–860
Uno H, Cai T, Pencina MJ, D'Agostino RB, Wei LJ (2011) On the C-statistics for evaluating overall adequacy of risk prediction procedures with censored survival data. Stat Med 30(10):1105–1117. doi:10.1002/sim.4154
Van Belle V, Pelckmans K, Van Huffel S, Suykens JA (2011a) Improved performance on high-dimensional survival data by application of survival-SVM. Bioinformatics 27(1):87–94. doi:btq617 [pii]10.1093/bioinformatics/btq617
Van Belle V, Pelckmans K, Van Huffel S, Suykens JA (2011b) Support vector methods for survival analysis: a comparison between ranking and regression approaches. Artif Intell Med 53(2):107–118. doi:S0933-3657(11)00076-5 [pii]10.1016/j.artmed.2011.06.006
van de Vijver MJ, He YD, van't Veer LJ, Dai H, Hart AA, Voskuil DW, Schreiber GJ, Peterse JL, Roberts C, Marton MJ, Parrish M, Atsma D, Witteveen A, Glas A, Delahaye L, van der Velde T, Bartelink H, Rodenhuis S, Rutgers ET, Friend SH, Bernards R (2002) A gene-expression signature as a predictor of survival in breast cancer. N Engl J Med 347(25):1999–2009
van der Laan MJ, Pollard KS, Bryan J (2003) A new partitioning around medoids algorithm. J Stat Comput Simulat 73(8):575–584
van Houwelingen H, Bruinsma T, Hart AA, van't Veer LJ, Wessels LFA (2006) Cross-validated Cox regression on microarray gene expression data. Stat Med 25:3201–3216
van Rijsbergen C (1979) Information retrieval, 2nd edn. Butterworths, London
van't Veer LJ, Dai H, van de Vijver MJ, He YD, Hart AA, Mao M, Peterse HL, van der Kooy K, Marton MJ, Witteveen AT, Schreiber GJ, Kerkhoven RM, Roberts C, Linsley PS, Bernards R, Friend SH (2002) Gene expression profiling predicts clinical outcome of breast cancer. Nature 415(6871):530–536
Verweij PJM, van Houwelingen JC (1993) Cross-validation in survival analysis. Stat Med 12:2305–2314
Wang Y, Klijn JGM, Zhang Y, Sieuwerts AM, Look MP, Yang F, Talantov D, Timmermans M, Meijer-van Gelder ME, Yu J, Jatkoe T, Berns EMJJ, Atkins D, Foekens JA (2005) Gene-expression profiles to predict distant metastasis of lymph-node-negative primary breast cancer. Lancet 365(9460):671–679. doi:10.1016/S0140-6736(05)17947-1
Webb A (2003) Statistical pattern recognition, 2nd edn. Wiley, New York
Wei JS, Greer BT, Westermann F, Steinberg SM, Son CG, Chen QR, Whiteford CC, Bilke S, Krasnoselsky AL, Cenacchi N, Catchpoole D, Berthold F, Schwab M, Khan J (2004) Prediction of clinical outcome using gene expression profiling and artificial neural networks for patients with neuroblastoma. Cancer Res 64(19):6883–6891. doi:64/19/6883 [pii]10.1158/0008-5472.CAN-04-0695
Weiss SM, Kulikowski CA (1991) Computer systems that learn. Morgan Kaufmann, San Mateo
Welford SM, Gregg J, Chen E, Garrison D, Sorensen PH, Denny CT, Nelson SF (1998) Detection of differentially expressed genes in primary tumor tissues using representational differences analysis coupled to microarray hybridization. Nucleic Acids Res 26(12):3059–3065
Wilson CL, Miller CJ (2005) Simpleaffy: a BioConductor package for affymetrix quality control and data analysis. Bioinformatics 21(18):3683–3685
Wu Z, Irizarry RA (2004) Preprocessing of oligonucleotide array data. Nat Biotechnol 22:656–658
Yeung KY, Bumgarner RE (2003) Multiclass classification of microarray data with repeated measurements: application to cancer. Genome Biol 4(12):R83. doi:10.1186/gb-2003-4-12-r83
Youden WJ (1950) Index for rating diagnostic tests. Cancer 3(1):32–35
Zhu J, Hastie T (2004) Classification of gene microarrays by penalized logistic regression. Biostatistics 5(3):427–443. doi:10.1093/biostatistics/5.3.427 5/3/427 [pii]
Zilliox MJ, Irizarry RA (2007) A gene expression bar code for microarray data. Nat Methods 4(11):911–913. doi:nmeth1102 [pii]10.1038/nmeth1102
Zou H, Hastie T (2005) Regularization and variable selection via the elastic net. J R Stat Soc SerB Stat Methodol 67(2):301–320. doi:10.1111/j.1467-9868.2005.00503.x

Chapter 12
Challenges in the Commercialization of Array-Based Tests

Justin Petrone

12.1 Introduction

Before we can discuss the challenges faced in commercializing array-based tests in depth, we have to get past the ambiguous language that is often applied to these high-complexity devices and the uses for which they are intended.

CGC Genetics, for instance, offers a menu of genetic tests it calls "Array CGCs," but these assays are actually run on Illumina's digital microbead-based BeadXpress platform and, therefore, are not array-based tests (Petrone 2010a). The term "test" or "diagnostic" is even more problematic, as many firms have managed to sell arrays with a research-use-only label to clinical geneticists who then use them to diagnose patients with suspected constitutional abnormalities. When the array leaves the factory, it is a "research device," but when it arrives in the customer's mailbox, it becomes a "test" (Petrone 2009a). This elegant dance between "research use" and "clinical use" is partially born from uncertainties in the regulatory landscape. As authorities work to find ways to regulate arrays without hindering their impact in treating patients, companies and users are also finding ways to use the latest technology in a clinical setting without breaking the law.

For our purposes, an array-based test here is defined as one that is run on a traditional microarray, that is, an array of biomolecules bound to surface or substrate. This excludes digital microbead-based platforms, such as the BeadXpress. The arrays that we will discuss are manufactured by a number of large vendors, including Agilent Technologies, Affymetrix, Illumina, and Roche NimbleGen, but also by other vendors, and used by pure-play molecular diagnostic companies like GenMark Diagnostics, AutoGenomics, SQI Diagnostics, and many others.

To date, only a handful of tests have managed to successfully navigate the process of attaining clearance as a medical device in the USA, while others are

J. Petrone (✉)
BioArray News, GenomeWeb LLC, New York, NY, USA
e-mail: jpetrone@genomeweb.com

B. Jordan (ed.), *Microarrays in Diagnostics and Biomarker Development*,
DOI 10.1007/978-3-642-28203-4_12, © Springer-Verlag Berlin Heidelberg 2012

offered in Europe and elsewhere under slightly less stringent regulatory conditions. Moreover, some diagnostics have been offered as laboratory-developed tests in US Clinical Laboratory Improvement Amendments-compliant facilities, as some companies prefer to offer array-based LDTs rather than taking their tests through the process of gaining regulatory clearance, especially in the American market.

This has resulted in two groups of array-based diagnostics: those that have gained US Food and Drug Administration clearance (or are being prepared with an FDA submission in mind) and those that are offered as LDTs in a CLIA-certified environment. As we will see later in the chapter, arrays have arguably had more success in terms of adoption as LDTs than FDA-cleared diagnostic kits. Still, the bulk of molecular diagnostics companies developing array-based tests today are doing so with the intent to obtain regulatory clearance for them.

Both categories of array-based tests—cleared devices and LDTs—face similar challenges in the process of reaching the market as well as in market acceptance. These include but are not limited to the amount of evidence that exists proving the scientific foundation of the given tests, high costs associated with the technology and reluctance from insurance companies to reimburse those costs, patients' right to privacy and confidentiality of results, awareness from the targeted medical community of the existence and benefit of these new tests, and, finally, a shifting regulatory landscape, where authorities are struggling continuously to regulate these complex tests without thwarting their ability to provide benefits for patients (Li et al. 2008).

This chapter will provide an overview of these issues by discussing some key examples of array-based tests and the challenges their developers experienced them in commercializing them. Tests that have achieved regulatory clearance for clinical use will be discussed first, while LDTs will be addressed later in the chapter. Finally, the challenges facing the commercialization of array-based tests will be summarized at the end of the chapter.

12.2 Key Examples of the Path to Regulatory Clearance

12.2.1 Roche's AmpliChip CYP450

The first array-based test to obtain US Food and Drug Administration clearance was Roche's AmpliChip CYP450 in December 2004. It is worth discussing the AmpliChip and its path to market in detail, because many of the expectations for this test and the challenges it has encountered in being commercialized have impacted and will impact other array-based tests being groomed for FDA clearance.

The AmpliChip CYP450 allows detection of gene variations—including deletions and duplications—for the CYP2D6 and CYP2C19 genes, which play a major role in the metabolism of an estimated 25% of all prescription drugs (Roche 2011). Specifically, the test detects up to 33 CYP2D6 alleles and 3 CYP2C19 alleles, as well as CYP2D6 gene duplication and deletions. Roche claims the test helps clinicians determine therapeutic strategy and treatment dose for therapeutics metabolized by the CYP2D6 or CYP2C19 gene product.

The AmpliChip CYP450 is manufactured by Santa Clara, Calif.-based Affymetrix in the cartridge format of the vendor's GeneChip platform and processed using its GeneChip Scanner 3000DX instrument. The relationship between Roche and Affymetrix set the precedent for similar agreements, dubbed "Powered by Affymetrix" partnerships, between Affymetrix and molecular diagnostics companies that wished to develop and launch tests on its platform. At the time, both Roche and Affymetrix were optimistic about commercializing the technology, with one Roche official at the time touting the duo as capable of "bring personalized medicine to fruition" (Krochmal 2003).

Roche was keen on seeing arrays adopted for clinical use. In January 2003, the Swiss diagnostics giant announced a $70 million, 18-year licensing agreement with Affymetrix ahead of launching the chip in June 2003, with the idea that Roche would launch other AmpliChips for other indications in the years ahead. The company initially marketed the AmpliChip CYP450 as an analyte-specific reagent, a category exempt from the FDA's 510(k) premarket notification requirements. ASRs are defined as reagents which, through specific binding or chemical reaction with substances in a specimen, are intended for use in a diagnostic application for identification and quantification of an individual chemical substance or ligand in biological specimens. ASRs therefore are considered to be part of a diagnostic test, but not the test itself (Shapiro and Prebula 2003).

This designation was short lived. In November 2003, the FDA decided that, based on review of the press release and other Roche materials, the AmpliChip was not a reagent, but a multisignal device, and one that was intended for use with a specific configuration of a proprietary accessory, the Affymetrix platform. The agency further determined that the product was subject to premarket review, meaning that it would have to be evaluated by the agency before it could be used in any clinic. In response to this change in the regulatory landscape, Roche then pledged to seek clearance for the AmpliChip CYP450 in Europe and the USA.

Three submission routes were available to Roche in the USA: premarket approval, a 510(k), or a de novo 510(k). The premarket approval process was and is the most stringent, as it typically covers life-sustaining devices, such as heart valves; a 510(k) path is intended for a device that is equivalent to an existing product, and a de novo 510(k) is intended for devices that do not have an equivalent and that do not merit PMA regulatory requirements. Ultimately, Roche and Affymetrix submitted the AmpliChip CYP450 as a de novo 510(k). After attaining a CE-IVD mark for the test in Europe in September 2004, the AmpliChip CYP450 and the Affymetrix GeneChip platform were submitted to US regulators in December 2004 and achieved clearance for clinical use later that month (Winnick 2004a).

Roche and Affymetrix's dialogue with the FDA revealed two immediate challenges for the commercialization of an array-based test. The first was the uncertain regulatory route for a new technology: Roche initially marketed the test as an ASR, only to later seek de novo 510(k) clearance for the AmpliChip CYP450 as a medical device. But connected with regulatory oversight was a delay to market.

It took a full 18 months from when Roche first launched the AmpliChip CYP450 as an ASR to when it achieved FDA clearance and could be sold to labs in the USA. A third challenge related to this process was the cost of providing the requisite data for FDA clearance, which to date is unknown.

Despite these challenges, the path to gaining FDA clearance was significant, in that it paved the way for similar array-based tests, especially those based on the Affymetrix platform. Other molecular diagnostic companies now had a route forward for getting their tests to market, and Affymetrix had a system that had been cleared for clinical use, a major sales point for the company as it defended its large stake in the array market from a growing number of competitors.

Though it took 18 months from its launch date to FDA clearance, Roche remained optimistic about the array-based test. The company officially said it was targeting AmpliChip CYP450 revenues of over $100 million by the year 2008 (Krochmal 2003). Another official said the test was targeting peak global sales of about $80 million, and said that by 2014, AmpliChip products would be competing in a market worth around $11 billion. The company also planned to expand its AmpliChip menu and had discussed tests in development for breast cancer classification, osteoporosis, leukemia, and the early detection and classification of prostate cancer (Winnick 2004a).

Yet upon achieving FDA clearance for the AmpliChip, Roche faced even more challenges in selling it. These included the cost of the test, gaining reimbursement for it, and educating target users about its clinical utility. The cost of the test was high, between $600 and $1,300, and the Affymetrix platform itself, roughly $160,000 at the time, was a significant capital investment for any lab. This arguably had an impact on acceptance. As of 2007, eight labs offered the AmpliChip CYP450 test in the USA (Allison 2008).

While some diagnostic labs were able to offer the test, health insurance companies were reluctant to reimburse it, citing questions about its utility. The fact that the test was suggested as a way to assess drug metabolism, rather than mandated by the FDA for use in certain situations, meant that clinicians wary of its costs were not obligated to use it. And just as cost and reimbursement emerged as major challenges to the success of this first array-based test, so did informing the market. As one Roche official pointed out in a 2008 *Nature Biotechnology* article, the psychiatry community, to whom the AmpliChip CYP450 was targeted, was not used to adopting new tests and also unaware of the test's benefits (Allison 2008).

One way Roche could have reached target customers was via papers describing the usefulness of the test in medical literature. But it took some time for publications about the AmpliChip CYP450 to appear. According to a search of PubMed.gov, between 2004 and 2011, 28 papers have been published that explicitly mention the AmpliChip CYP450. All of them were published after the test's launch though, and the majority, 16 out of 28, were published in the last 2 years, 5 years after the FDA cleared the test.

Despite these challenges, response to the test has been mostly favorable, with a few criticisms. For instance, a 2009 review of the test in psychiatric practice determined that the AmpliChip CYP450 played a "supporting role with regard to dosing decisions, rather than being the main influence on clinician behavior" and said that the test was viewed as "potentially useful in facilitating the development of the doctor–patient relationship." Reported disadvantages of the test included "cost and challenges inherent in introducing a new clinical procedure into routine practice." The reviewers concluded that "psychiatric clinicians are receptive to employing this test as a clinical support tool if its implementation is carefully considered and economically justifiable" (Dunbar 2009).

It is unclear if Roche ever met its mark of generating $100 million in AmpliChip sales by 2008. The firm does not break out its sales by product. But one demonstration of the challenges facing adoption of array-based tests is the fact that the company has not since 2004 taken another AmpliChip through the 510(k) submission process, in spite of earlier pledges to launch multiple array-based tests.

12.2.2 The AmpliChip p53

Of the other AmpliChips developed by Roche, the AmpliChip p53 Research Test is arguably the second most widely known. The test, also based on the Affymetrix platform, is designed to detect single-nucleotide substitutions and 1-base pair deletions in the entire coding region and the flanking splice sites of exons 2–11 of the p53 gene. Users of the test can detect the presence of sequence alterations through comparative analysis of the hybridization pattern of a series of probes to sample DNA and wild-type reference DNA. Roche has claimed that the test can potentially lead to better treatment outcomes for cancer patients by determining which patients would most likely respond to investigational drug candidates.

Rather than take the AmpliChip p53 through the FDA clearance process, Roche has instead decided to offer the chip as a research tool to be used in developing drugs. In 2010, the firm announced a deal with Merck, whereby Merck gained the right to use the AmpliChip p53 to discover and develop innovative cancer therapeutics. Specifically, Merck aimed to apply the AmpliChip p53 test in selected clinical trials to identify those patients most likely to respond to specific therapeutic regimens in development. Until it inked the deal with Merck, Roche had solely used AmpliChip p53 internally. Roche said at the time that the AmpliChip p53 was "intended for investigational use as a companion diagnostic test in cancer clinical trials" and that the company has not yet decided to submit the test to the FDA for clearance (Petrone 2010b).

In the meantime, a body of literature is being built around the test. Since 2010, three papers have appeared concerning the use of the AmpliChip p53 to predict tumor response in patients being treated for breast cancer and chronic lymphocytic leukemia (Glück et al. 2011; Chiaretti et al. 2011; Baker et al. 2010). Roche therefore still sees value in using and developing array-based tests but has, at least in the case of the AmpliChip p53, shifted its commercialization strategy.

12.2.3 The "Powered by Affymetrix" Program

Following the FDA's clearance of the AmpliChip CYP50, Affymetrix sought to add new partners to its "Powered by Affymetrix" program. The program consisted of firms that planned to develop and launch tests on its GeneChip platform.

The roster of diagnostics partners bore names like Roche, BioMerieux, and J&J unit Veridex, Pathwork Diagnostics, Ipsogen, and Almac Diagnostics. As of August 2011, Affymetrix's website lists 12 of such partners: Almac, BioMerieux, Epigenomics, Ipsogen, Medical Prognosis Institute, Pathwork Diagnostics, Roche, Skyline Diagnostics, Sysmex, TessArae, Veridex, and Vita Genomics (Affymetrix 2011). Other companies have also publicly discussed plans to launch tests based on the Affymetrix GeneChip platform, such as Signal Genetics, ChipDx, and Allegro Diagnostics. The total number of firms and organizations developing Affymetrix-manufactured tests is unclear.

What is clear is that of the more than dozen firms that have initiated test development programs on the Affymetrix platform since the AmpliChip was cleared by the FDA in 2004, only one has received similar clearance: Pathwork Diagnostics.

Founded in 2004, Redwood City, Calif.-based Pathwork's flagship diagnostic is its Tissue of Origin Test. The Affymetrix-manufactured diagnostic compares the expression of more than 1,500 genes in a tumor sample against 15 known tissues that represent more than 60 morphologies in order to identify the source of cancers that are difficult to classify and increase oncologists' confidence in their treatment approaches. Pathwork claims the test can provide an objective, probability-based score for each potential malignancy. The test is meant to be used with the patient's clinical history and complementary diagnostics, such as immunohistochemistry (Petrone 2010c).

As with the AmpliChip CYP450, Pathwork's road to regulatory clearance for the TOO Test took time. The firm initially aimed for a 2006 launch date, but did not begin offering the test until 2 years later, launching it as an LDT through its CLIA-certified laboratory in April 2008 until it received FDA clearance for the test in August of that year. Even upon achieving clearance, the TOO Test could only be used with frozen tissue samples until the firm achieved a second clearance for use with formalin-fixed, paraffin-embedded samples in 2010.

This is an example of a challenge facing all companies developing array-based tests with FDA clearance in mind: the tests are taking longer than expected to reach the market. That in turn has an impact on the expectations for array-based diagnostics. The days of predicting millions or billions of dollars in revenue for a particular test, as was done with the AmpliChip CYP450 marketing, are over. It also creates challenges for small diagnostic companies in terms of remaining financially solvent long enough to see a test reach market and begin to return some of that investment. Developing and seeing a test make it to the clinic therefore requires a significant financial resources.

Since its founding, Pathwork has raised $62 million in three financing rounds to fund the development and commercialization of the TOO Test and others in its pipeline. At the same time, in those 7 years, Pathwork has managed to launch just one test. It is unclear how successful the TOO Test has been as Pathwork is a privately held company, but Pathwork has sought to overcome some of the headwinds faced by Roche's CYP450 AmpliChip.

It moved quickly to introduce the test to target customers via the medical literature. In January 2008, ahead of the TOO Test's launch via Pathwork's CLIA lab, the firm saw the test discussed in the *Journal of Molecular Diagnostics* (Dumur et al. 2008). Since then, nine more papers have discussed the test, most recently one describing a validation study of the assay in FFPE samples, which appeared in January 2011 in the same journal (Pillai et al. 2011).

Pathwork also worked to set reimbursement schemes for the TOO Test. So far, the company has won Medicare coverage and is "working with insurance companies across the country to secure coverage for the test," according to the firm's website. It also offers financial assistance to customers that cannot afford to pay for the test.

Of the other firms developing tests on the Affymetrix platform, perhaps Skyline Diagnostics is the closest to seeing its assay cleared by the FDA. In March 2011, the Dutch company received a CE-IVD mark for AMLprofiler, a microarray-based assay for acute myeloid leukemia. The certification allows the Rotterdam-based firm to offer AMLprofiler to clinicians within the European Union. At the time of gaining the CE-IVD mark, Skyline said it was on track to achieve FDA premarket approval for the test in 2012 (Petrone 2011a).

AMLprofiler is an in vitro diagnostic that Skyline claims simplifies the recognition of AML subtypes by replacing with one test the approximately eight separate assays currently used by clinicians to monitor the disease. Using AMLprofiler, clinicians can reclassify a significant portion of the intermediate-risk group of AML patients into either the favorable or the unfavorable risk categories, "facilitating and improving the clinical decision-making process," Skyline claims. The firm offers its test via its CLIA-compliant lab as well as a kit. The ability for customers to have the test processed by Skyline enables the firm to reduce the cost for the test, as there is no need for a lab to make capital expenditures to offer an array-based diagnostic.

Still, while Pathwork and Skyline show that Affymetrix's diagnostic partners will continue to bring tests to market in the USA as cleared medical devices, some of its partners have opted to move their assays to other technology platforms, citing a number of factors.

Marseille, France-based Ipsogen, for instance, originally launched in the EU its Affymetrix-manufactured MapQuant Dx Genomic Grade, which uses a 97-gene signature taken from biopsies to measure tumor grade—a consensus indicator of tumor proliferation, risk of metastasis, and response to chemotherapy. However, the company has recently opted to move the test to a quantitative PCR platform instead.

Berlin-based Epigenomics made a similar decision. The company originally developed an assay on the Affymetrix platform for PITX2 DNA methylation, an independent prognostic factor in prostate and breast cancer. In 2011, the firm opted to move the test to RT-PCR, citing the low number of markers, as well as RT-PCR's "multiplexing capabilities, widespread use and acceptance, availability of numerous platforms with significant installed bases, ease of use, throughput, and reagent costs" (Petrone 2011b).

BioMérieux has also decided against launching array-based tests for similar reasons. Marcy l'Etoile-based BioMérieux had signed numerous agreements over the past decade to develop and market tests on the GeneChip platform, and assays for prostate cancer, breast cancer, sepsis, and infectious diseases had been discussed. The company recently made a strategic decision though to transfer all of its array-based assays in development to other systems, such as Biocartis' DNA/RNA molecular diagnostics platform (Petrone 2011c).

The availability of more widely accepted technology platforms is therefore another challenge facing the commercialization of array-based tests.

12.2.4 Agendia's MammaPrint

The decision of some molecular diagnostics players who had initially developed array-based tests to transfer those assays to RT-PCR is a good entry point for discussing the case of Agendia's MammaPrint test and highlighting a way in which competitive technologies create challenges for the acceptance of array-based tests.

Agendia was founded in 2003 as a spin-off from the Netherlands Cancer Institute and Antoni van Leeuwenhoek Hospital in Amsterdam to commercialize array technology for cancer diagnosis and drug development. In 2004, it launched the initial version of its MammaPrint breast cancer recurrence test in Europe. In 2006, the test was subjected to an international independent retrospective validation which led to the selection of MammaPrint as the stratification tool for the Microarray In Node-negative and 1 to 3 positive lymph node Disease may Avoid ChemoTherapy, or MINDACT trial, that featured 6,000 patients at 109 institutions in nine European countries (Agendia 2011a).

In 2007, after several years of discussion, Agendia submitted MammaPrint to the FDA for clearance, and the test was, after evaluation, cleared as an in vitro diagnostic multivariate index assay, a new test category defined in a series of draft guidances that were written in response to the number of high-complexity tests in development.

As the FDA defined it at the time, an IVDMIA is a device that combines the values of multiple variables using an interpretative function, such as a bioinformatics algorithm, to provide a single, patient-specific result that is intended for use in the diagnosis of disease, and which provides a result whose derivation is nontransparent and for this reason cannot be independently derived or verified by the end user (College of American Pathologists 2010). MammaPrint was the first test to be cleared as such following a 510(k) submission. Pathwork's TOO Test was also cleared as an IVDMIA a year later.

MammaPrint relies on a 70-gene expression signature to identify which early-stage breast cancer patients are at risk of distant recurrence following surgery, independent of estrogen receptor status and any prior treatment. The test is used to stratify patients into two distinct groups—low risk or high risk of distant recurrence. These results can have a direct impact on breast cancer treatment. As the firm states on its website, "hormonal therapy alone . . . may be sufficient to further reduce her risk if your patient is Low Risk by MammaPrint, when combined with traditional risk factors" while a High Risk result may point toward "more aggressive therapy including chemotherapy."

Agendia's test is manufactured by Agilent Technologies and is offered via CLIA-compliant facilities in the Netherlands and the USA.

Some could argue that Agendia did most things right in developing and launching MammaPrint. For one, the company worked to raise its profile in the medical literature. A search of PubMed.gov, for example, reveals that as of August 2011, 60 papers have been published that explicitly mention MammaPrint, seven of which were published in the year of the test's US launch. The test was selected for use in another large, multicenter trial in the USA, the National Cancer Institute supported I-SPY I and I-SPY II breast cancer clinical trials which focus on the prediction of therapeutic response in the neoadjuvant setting. Finally, Agendia moved quickly to set up reimbursement schemes for the test's $4,200 US price tag. In 2009, the company gained Medicare coverage for the test and bills private health insurance companies and third-party payors on behalf of insured patients throughout the USA, according to the firm's website.

Over 3 years, Agendia established for MammaPrint US regulatory clearance, awareness, and reimbursement. The greatest challenge for the company in commercializing its test though has been a competitive assay offered on a more widely accepted technology platform: Genomic Health's RT-PCR-based Oncotype Dx.

Oncotype DX quantifies the likelihood of disease recurrence in women with early-stage hormone estrogen receptor positive only breast cancer and assesses the likely benefit from certain types of chemotherapy. The test, which is offered as an LDT through Genomic Health's CLIA-compliant lab, relies on a panel of 21 genes to determine a Recurrence Score, a number between 0 and 100 that corresponds to a specific likelihood of breast cancer recurrence within 10 years. Additionally, the test can be used to stratify patients into low-, intermediate-, or high-risk categories when the patient receives 5 years of tamoxifen (Genomic Health 2011a). The test costs roughly the same amount as MammaPrint.

When compared to MammaPrint, Oncotype Dx though had several market advantages. It became available in 2004, 3 years before MammaPrint, giving it first-mover status in the market. The test was also less complex, relying on a panel of 21 genes versus 70 genes, and on a technology platform, RT-PCR, that users were more familiar with. Also, Genomic Health, based in Redwood City, Calif., had the USA as its home market, whereas it took Agendia time to build a US presence. For its first 2 years on the market, MammaPrint tests were processed in Amsterdam before Agendia established a CLIA-compliant lab in California. Finally, Genomic Health was able to accept FFPE samples, while Agendia could accept only frozen tissue samples.

Still, Agendia also faced questions about the performance of its test. In 2011, the 12th St. Gallen International Breast Cancer Conference Expert Panel, for example, agreed that Oncotype DX could be used where available to predict chemotherapy responsiveness in an endocrine responsive cohort where uncertainty remained after consideration of other tests. In the same report, the panel agreed that the chemopredictive properties of MammaPrint "were not yet sufficiently established" (Goldhirsch et al. 2011).

The firm cited the lack of winning inclusion in "influential clinical guidelines," such as the St. Gallen guidelines, as a risk in its initial public offering prospectus, filed in June 2011. Agendia later withdrew its IPO, through which it had hoped to earn up to $75 million, the same month, citing unfavorable market conditions.

It is unclear how many tests Agendia has run since MammaPrint was launched. The company offers a number of other tests. According to the company's prospectus (Agendia 2011b), the firm's revenues were $4.7 million in 2010, more than tripling its 2009 sales of $1.4 million, and dwarfing its 2008 revenues of $500,000.

Genomic Health has sold more tests. It claims that its test has been used to guide treatment for 200,000 patients in over 60 countries. The company reported total 2010 revenue of $178.1 million, up 19% from $149.5 million in 2009 (Genomic Health (2011b)).

12.3 Other Array-Based Tests that Have Won Clearance

While attention has been paid to Roche's AmpliChip CYP450, Pathwork Diagnostics' TOO Test, and Agendia's MammaPrint, a number of lower-complexity, array-based tests have managed to win FDA clearance in recent years. Though these tests have not won their developers' revenues on par with Genomic Health, they have generated continuous growth for those companies. Also, these companies have accomplished this growth not by seeking to create new markets with their tests, but by looking to replace older technologies with more efficient, multiplexed tests based on low-density arrays.

An example of such a company is Pasadena, Calif.-based GenMark Diagnostics. Formerly known as Osmetech Molecular Diagnostics, GenMark has had three tests cleared by the FDA in recent years: its Cystic Fibrosis Genotyping Test, which detects preconception risk of cystic fibrosis; its Warfarin Sensitivity Test, which determines an individual's ability to metabolize the oral anticoagulant warfarin; and its Thrombophilia Risk Test, which detects increased risk of blood clots. The firm is also in negotiations with the agency regarding a number of other tests (GenMark Diagnostics 2011).

GenMark's tests are run in its eSensor XT-8 System. The firm's technology is based on the electrochemical detection of targets on a microarray. Using the benchtop XT-8, customers can process 24 test cartridges at a time.

GenMark has seen demand for its tests and system rise in recent years. In the second quarter of 2011, for example, total revenues rose 35% to $901,000 from $665,000 in 2010. The rise in growth has been attributed to growth in menu and systems placements but also to increased sales and marketing activities.

Another example of a company that has achieved FDA clearance is Carlsbad, Calif.-based AutoGenomics has had five assays cleared by the agency: for Factor II, Factor V Leiden, Factor II–V Leiden, Warfarin, and CYP2C19. The assays run on its Infinity Analyzer system, an automated, benchtop system that processes AutoGenomics' BioFilmChip arrays using fluorescence-based detection. AutoGenomics is privately held, and so its sales are unknown. It continues to introduce new tests on its system (Autogenomics 2011).

Finally, Toronto-based SQI Diagnostics has had two panels cleared by the FDA. Its IgXplex celiac qualitative assay, a 4-plex microarray-based test for antitissue transglutaminase IgG and IgA, was cleared in 2011. SQI's IgXplex rheumatoid arthritis assay was cleared by the agency in late 2009.

Both had a relatively quick route to market. SQI was founded in 2006, and its platforms consist of the automated SQiDWorks system and SqiDman, a semiautomated version of SQiDWorks for lower-throughput users. The company also sells the QuantiSpot, a 96-well microarray consumable. The SQiDWorks and SqiDman systems can typically process multiplexed assays to quantify serum concentrations of up to 12 individual biomarkers, or qualitatively detect up to 24 biomarkers per patient, according to SQI.

SQI filed for an initial public offering in the USA in 2011. The company's revenues for its fiscal year ended Sept. 30, 2010 were $36,500 in total revenues, including $5,000 in product revenues. For the first half of fiscal 2011, ended March 31, 2011, it recorded $23,000 in revenues, including $13,000 in product revenues (SQI Diagnostics 2011). SQI's sales are therefore growing, although they remain relatively small.

What do GenMark, AutoGenomics, and SQI have in common? They all have built their businesses around a model of placing automated benchtop systems with customers, typically clinical or reference laboratories, and growing product sales by introducing new tests.

In the case of GenMark, sales have been directly attributable to sales and marketing activities. But most of all, these FDA-cleared tests are low complexity, in that they test only a handful of markers per assay, and they replace existing tests on the market that are run on lower-throughput technologies, such as RT-PCR, Sanger sequencing, or enzyme-linked immunosorbent assays. Therefore, there are fewer technological and scientific questions about the tests. This recipe has proved successful for these firms, in spite of the challenges facing their newer technologies, and resulted in increasing, albeit modest, revenue numbers.

12.4 A Highly Successful Entry: LDTs and Array-Based Cytogenetics

As it has for GenMark Diagnostics, intensified sales and marketing activities have also paid off for Irvine, Calif.-based CombiMatrix Diagnostics. The firm performs DNA-based testing services for cancer and developmental disorders using assays that are run on Agilent Technologies-manufactured chips or on internally printed bacterial artificial chromosome arrays. In recent quarters, its revenues have consistently grown. In the second quarter of 2011, for instance, its revenues increased 32% to $1.2 million from $916,000 in the second quarter of 2010. For the 6 months ended June 30, 2011, total revenues increased by 22% to $2.1 million as compared to $1.7 million for the comparable 2010 period (CombiMatrix 2011b).

The company has attributed its revenue increases directly to investments in sales and marketing (Petrone 2011d). But CombiMatrix has also gone a different route in bringing its diagnostics to market. It has chosen to offer them solely in a CLIA-certified environment as laboratory-developed tests. This has allowed the firm to avoid spending money to generate the data required for FDA clearance and also to more rapidly introduce new tests. The company also sells a service rather than a system and reagents, forgoing the "razor-razorblade" sales model that has largely dominated the market when it comes to FDA-cleared array-based tests.

The rationale for CombiMatrix's decision to offer its tests solely as LDTs is also related to the kinds of tests it offers. Its array comparative genomic hybridization (CGH)-based approach to detecting the genetic sources of developmental disorders in both prenatal and postnatal patients relies on comparing an affected individual's genome against a unaffected one in order to detect the genetic differences that may be causing the affected individual's phenotype.

Rather than surveying a sample for certain markers that have been proven to be correlated with cystic fibrosis or drug metabolism, CombiMatrix's geneticists therefore are instead looking across a genome for copy number variants or insertions/deletions that may be associated with a particular disorder. Often they find differences that are strongly associated with a certain disease or disorder. Sometimes they are confronted with inconclusive findings that may only become meaningful at a later date following new discoveries. Needless to say, it would be difficult to have such an assay cleared by the FDA. If only a few people in the world suffer from a certain genetic disorder, companies like CombiMatrix would likely find it difficult to validate the ability of their array-based test to detect that disorder in a statistically meaningful study.

CombiMatrix is just one example of a larger phenomenon of array adoption in cytogenetics. For nearly a decade, geneticists at companies and institutions have been adopting so-called chromosomal arrays for detecting developmental disorders, with all of the testing has taken place within a CLIA-compliant environment.

The ability to operate in a CLIA setting benefitted companies and labs offering such services in several ways when compared to companies and labs that wished to sell FDA-cleared diagnostic kits. First, these companies and labs did not have to take their tests through the regulatory clearance process, saving them time and logistical and financial resources. Second, since they were not under the same scrutiny as companies seeking clearance, so these companies and labs were able to quickly update their tests based on new findings or the availability of better technologies, each time without having to go through a rigorous evaluation process.

At the same time, companies and labs using chromosomal arrays have sought to set standards for the kinds of arrays they use and the manner in which they interpret and report back their findings to clinicians or patients. One organization that has played a role in setting standards for the community of array users is the International Standards for Cytogenomic Arrays Consortium, which was founded to arrive at a consensus design for the chips used to test patients, as well as to set standards for data interpretation and sharing.

ISCA has worked together to press organizations such as the American Society for Medical Genetics to recommend array-based methods as a replacement to older technologies like karyotyping and fluorescence in situ hybridization as first-tier tests in detecting developmental abnormalities. The ACMG issued such guidelines in 2010, making arrays the standard of care for detecting developmental disorders in the USA (Manning et al. 2010). More detailed recommendations for array design and data interpretation were published in 2011 (Kearney et al. 2011).

Companies have looked to profit from this sea change in the cytogenetics tools market. Agilent Technologies has acted as the foundry for several companies that sell arrays specifically for cytogeneticists, such as Oxford, UK-based Oxford Gene Technology and Cambridge, UK-based BlueGnome. Affymetrix, Roche NimbleGen, and Illumina have also developed specific offerings to sell to geneticists. As noted at the beginning of this chapter, arrays are officially marketed as being for research use, and most cytogenetics labs confirm array-generated findings with an alternate technology, usually FISH.

Just as array vendors have profited from adoption of arrays by cytogeneticists, companies and labs that offer services have as well. Spokane, Wash.-based Signature Genomics, now part of PerkinElmer, has offered cytogenetic testing services since 2003, originally running samples on internally printed BAC arrays before graduating to Agilent-made oligo arrays, and later to Roche NimbleGen chips. Roche NimbleGen's current CGX array offering was designed by Signature (Petrone 2009b).

According to Signature, it began offering testing in March 2004 and, by August of that year, was making a profit. As of 2011, Signature has over 115 employees and has processed over 45,000 cases since its inception. PerkinElmer acquired the array-based testing firm in 2010 for $90 million. At the time of the acquisition, the firm was generating at least $10 million a year (Petrone 2010d).

Companies and labs like Signature have faced the same challenges in commercializing array-based tests as those who have sought FDA clearance for their diagnostics. These include changes in the regulatory environment, questions about cost and reimbursement, skepticism from customers about new technologies, and education and awareness issues. Signature Genomics and others have managed to overcome some of these challenges because they are offering LDTs, but are still affected by them.

Regulatory oversight of chromosomal array tests continues to evolve. For much of the past decade, companies and organizations that offer such tests have benefitted from uncertainty at the FDA about how the tests should be regulated. As previously noted, this has allowed companies like Signature relative freedom in updating tests and adding new array-based diagnostics to menu while avoiding a long test submission process.

At the same time, the FDA has since 2009 made it clear that it intends to eventually require array manufacturers to submit their chromosomal arrays to the agency for clearance. Questions about how the arrays will be assessed remain however unanswered. The agency has acknowledged that array-based cytogenetic tests have challenged its regulation strategy because they provide an "undefined number of outputs open to interpretation," as opposed to a set of defined outputs, as is the case with the array-based tests that the agency cleared (Petrone 2010e).

Some states, such as New York, already require companies and labs to submit their assays for review. In the event that a particular state seeks to evaluate a chromosomal array, it can take years for a company or lab to gain clearance. Such was the case with CombiMatrix. It took roughly 2 years for the company to gain a clinical laboratory permit from New York State for its postnatal DNAarray test (CombiMatrix 2011a).

In the meantime, reimbursement schemes for the test have become more apparent to the companies and labs offering them. CombiMatrix, for instance, bills its tests by multiplying certain Current Procedural Terminology codes. These are CT codes that exist for genetic testing, but predate the advent of array-based analysis. For one of its chromosomal array tests, CombiMatrix bills 88391 twice and 88386 five times. Signature Genomics meantime bills according to the kind of patient sample it receives. For a chromosomal array run on a postnatal peripheral blood sample, the firm bills 83891 once and 88386 six times.

In light of the discussion of how best chromosomal arrays should be regulated, vendors and users have argued for some time that the arrays should be evaluated on their ability to accurately measure copy number variation rather that detect certain syndromes, and that results should be interpreted by certified professionals.

Baylor College of Medicine's Arthur Beaudet, for instance, has likened arrays to magnetic resonance imaging machines and argued that the clinical interpretation of data generated by arrays falls within the practice of medicine and, thus, outside the FDA's regulatory purview (Schaaf et al. 2011).

Beaudet has also called the use of chromosomal arrays the "most significant clinical benefit that has come out of the Human Genome Project" (Schaaf et al. 2011) and has cautioned against forcing manufacturers through a clearance process that will limit the use of the technology, though he acknowledged that the current framework, where arrays are still offered as laboratory-developed tests, was unlikely to survive. As of September 2011, companies and labs that offer chromosomal arrays do so under existing regulatory conditions.

12.5 Conclusion

In the two decades since microarrays were first developed, the technology has gone from being a crude, research tool to the platform of choice for tests that were supposed to herald a "new age of personalized medicine" and earn their developers millions of dollars in revenues. As a 2010 discussion of the clinical use of arrays in *Expert Reviews in Molecular Diagnostics* summed up, "In the early 2000s, clinical microarrays were widely expected to represent a billion-plus market by 2009."

While some arrays have been adapted for clinical use, the "impact of microarrays in molecular diagnostics has remained relatively limited," and their market is currently estimated to be between $300 and $400 million, less than 10% of the total molecular diagnostics market (Jordan 2010).

As of 2011, only a handful of tests have been cleared for clinical use by US regulators. While some are generating consistent growth, their uptake has been dwarfed not only by molecular diagnostics firms' initial expectations but also by other molecular genetic tests on more widely accepted technology platforms.

And yet array-based tests have been indisputably successful. One area where they have risen to the forefront of genetic testing is clinical cytogenetics. While assays cleared by US regulators for clinical use have found a slow road to acceptance, the past few years have seen these laboratory-developed chromosomal array tests become the first-tier, standard of care tools for diagnosing developmental disorders in the postnatal setting.

The road for array-based tests from conception to market acceptance therefore has not been easy. At the same time, companies developing such tests have become more aware of the challenges facing their diagnostics and have worked to overcome them to the point that they have finally begun to win medical acceptance and generate revenues.

While arrays are not expected to dominate the molecular diagnostics market, they should be able to capture a sizeable share, perhaps as much as 20% of the market, which could reach $10 billion by 2015 (Jordan 2010). Experts believe that success will depend on technical improvements facilitating their use in a medical setting as well as on reasonable regulation, though it will be constrained by competition from other technological approaches.

References

Affymetrix (2011) Current powered by Affymetrix partners. http://www.affymetrix.com/partners_programs/pba_partners/index.affx#1_4

Agendia (2011a) A history of science. http://www.agendia.com/pages/history/87.php

Agendia (2011b) Agendia prospectus. http://www.afm.nl/registers/emissies_documents/6767.pdf

Allison M (2008) Is personalized medicine finally arriving? Nat Biotechnol 26:509–517

Autogenomics (2011) AutoGenomics molecular diagnostics products. http://autogenomics.com/products.php

Baker L et al (2010) p53 mutation, deprivation and poor prognosis in primary breast cancer. Br J Cancer 102(4):719–26

Chiaretti S et al (2011) Evaluation of TP53 mutations with the AmpliChip p53 research test in chronic lymphocytic leukemia: correlation with clinical outcome and gene expression profiling. Genes Chromosomes Cancer 50(4):263–74

College of American Pathologists (2010) In vitro diagnostic multivariate index assays. http://www.cap.org/apps/cap.portal?_nfpb=true&cntvwrPtlt_actionOverride=%2Fportlets%2FcontentViewer%2Fshow&_windowLabel=cntvwrPtlt&cntvwrPtlt{actionForm.contentReference}=committees%2Ftechnology%2Fivdmia.html&_state=maximized&_pageLabel=cntvwr

CombiMatrix (2011a) CombiMatrix receives New York State Clinical Laboratory Permit. http://investor.combimatrix.com/releasedetail.cfm?ReleaseID=597407

CombiMatrix (2011b) Combimatrix Corporation reports second quarter 2011 financial results. http://investor.combimatrix.com/releasedetail.cfm?ReleaseID=598171

Dumur C et al (2008) Interlaboratory performance of a microarray-based gene expression test to determine tissue of origin in poorly differentiated and undifferentiated cancers. J Mol Diagn 10 (1):67–77

Dunbar L (2009) Clinician experiences of employing the AmpliChip(R) CYP450 test in routine psychiatric practice. J Psychopharmacol. doi:10.1177/0269881109106957

GenMark Diagnostics (2011) GenMark diagnostics reports second quarter 2011 results. http://ir.genmarkdx.com/releasedetail.cfm?ReleaseID=597110

Genomic Health (2011a) The OncoType Dx Breast Cancer Assay. http://www.oncotypedx.com/en-US/Breast.aspx

Genomic Health (2011b) Genomic Health announces year-end 2010 financial results, provides 2011 financial outlook. http://investor.genomichealth.com/releasedetail.cfm?ReleaseID=548542

Glück S et al (2011) TP53 genomics predict higher clinical and pathologic tumor response in operable early-stage breast cancer treated with docetaxel-capecitabine ± trastuzumab. Breast Cancer Res Treat [Epub ahead of print]

Goldhirsch A et al (2011) Strategies for subtypes–dealing with the diversity of breast cancer: highlights of the St Gallen International Expert Consensus on the Primary Therapy of Early Breast Cancer 2011. Ann Oncol 22(8):1736–47

Jordan B (2010) Is there a niche for DNA microarrays in molecular diagnostics? Expert Rev Mol Diagn 10(7):875–82

Kearney H et al (2011) American College of Medical Genetics standards and guidelines for interpretation and reporting of postnatal constitutional copy number variants. Genet Med 13(7):680–5

Krochmal M (2003) Roche's AmpliChip Microarray faces regulatory hurdles for Tech and Tests. BioArray News. http://www.genomeweb.com/arrays/roche-s-amplichip-microarray-faces-regulatory-hurdles-tech-and-tests

Li X et al (2008) Clinical utility of microarrays: current status, existing challenges and future outlook. Curr Genomics 9(7):466–74

Manning M et al (2010) Array-based technology and recommendations for utilization in medical genetics practice for detection of chromosomal abnormalities. Genet Med 12(11):742–5

Petrone J (2009a) FDA may increase oversight over array-based clinical cytogenetic tests. BioArray News. http://www.genomeweb.com/arrays/fda-may-increase-oversight-over-array-based-clinical-cytogenetic-tests

Petrone J (2009b). Signature genomic labs to solely offer array CGH services on Roche NimbleGen Platform. BioArray News. http://www.genomeweb.com/arrays/signature-genomic-labs-solely-offer-array-cgh-services-roche-nimblegen-platform

Petrone J (2010a) CGC genetics adopts Illumina BeadXpress for clinical genetic testing. BioArray News. http://www.genomeweb.com/arrays/cgc-genetics-adopts-illumina-beadxpress-clinical-genetic-testing

Petrone J (2010b) Merck to use Roche's p53 AmpliChip for cancer research. BioArray News. http://www.genomeweb.com/arrays/merck-use-roches-p53-amplichip-cancer-research

Petrone J (2010c) FDA clears pathwork diagnostics' tissue-of-origin test for use with FFPE samples. BioArray News. http://www.genomeweb.com/arrays/fda-clears-pathwork-diagnostics-tissue-origin-test-use-ffpe-samples

Petrone J (2010d) PerkinElmer to acquire Cyto Array firm signature genomic labs for $90 M to Bolster Testing Services. BioArray News. http://www.genomeweb.com/arrays/perkinelmer-acquire-cyto-array-firm-signature-genomic-labs-90m-bolster-testing-s

Petrone J (2010e) FDA, labs, vendors take steps toward regulatory ground rules for array-based cytogenetic testing. BioArray News. http://www.genomeweb.com/arrays/fda-labs-vendors-take-steps-toward-regulatory-ground-rules-array-based-cytogenet

Petrone J (2011a) Skyline Diagnostics Preps Leukemia test for US launch; gains CE-IVD Mark. BioArray News. http://www.genomeweb.com/arrays/skyline-diagnostics-preps-leukemia-test-us-launch-gains-ce-ivd-mark

Petrone J (2011b). Despite years of development, Affymetrix's Dx partners slow to bring tests to market. http://www.genomeweb.com/arrays/despite-years-development-affymetrixs-dx-partners-slow-bring-tests-market

Petrone J (2011c) BioMerieux to move array-based tests to Biocartis, other MDx platforms. BioArray News. http://www.genomeweb.com/arrays/biomerieux-move-array-based-tests-biocartis-other-mdx-platforms

Petrone J (2011d) CombiMatrix to introduce gene-mutation testing; Q2 revenues Climb 31 percent. BioArray News. http://www.genomeweb.com/arrays/combimatrix-introduce-gene-mutation-testing-q2-revenues-climb-31-percent

Petrone J (2011e) Validation and reproducibility of a microarray-based gene expression test for tumor identification in formalin-fixed, paraffin-embedded specimens. J Mol Diagn 13(1):48–56

Pillai R et al (2011) Validation and reproducibility of a microarray-based gene expression test for tumor identification in formalin-fixed, paraffin-embedded specimens. J Mol Diagn 13(1):48–56

Roche (2011) AmpliChip CYP4450 Test. http://www.roche.com/products/product-details.htm?type=product&id=17

Schaaf C et al (2011) Copy number and SNP arrays in clinical diagnostics. Annu Rev Genomics Hum Genet 12:25–51 (Review)

Shapiro J, Prebula R (2003) FDA's regulation of analyte-specific reagents. Medical Device and Diagnostic Industry. http://www.mddionline.com/article/fda%E2%80%99s-regulation-analyte-specific-reagents

SQI Diagnostics (2011) SQI Diagnostics reports third quarter financial results. http://sqidiagnostics.com/about/press/2011/pr-08-29-11.php

Winnick E (2004a) Roche and Affy win EU approval for AmpliChip, Microarray Dx Platform. http://www.genomeweb.com/arrays/roche-and-affy-win-eu-approval-amplichip-microarray-dx-platform

Winnick E (2004b) Affymetrix, Roche file for US clearance of microarray-based diagnostic products. BioArray News. http://www.genomeweb.com/arrays/affymetrix-roche-file-us-clearance-microarray-based-diagnostic-products

Chapter 13
Genes, Microarrays, and Patents

Justin Petrone

13.1 Introduction

Gene patenting continues to be controversial. It merited a special plenary debate at the 2011 annual meeting of the American Society of Human Genetics conference in Montreal, where it was claimed that approximately 20% of the human genome is patented, but it was also noted that the area of genetic intellectual property continues to be in a state of flux. Experts tend to agree that protecting intellectual property is critical to encourage innovation, but question both the legitimacy of gene patents and whether or not their existence hinders the translation of discoveries into medical practice.

Microarrays are often cited as a research or diagnostics tool that could suffer from the patenting of individual genes. If 20% of the genome is patented, manufacturers or users would have to obtain licenses for all the genes they wished to include on any particular array, making such an endeavor cost prohibitive or the product unprofitable (SACGHS 2010). But to date, most of the litigation related to gene patents has impacted companies and laboratories that offer Sanger sequencing- or PCR-based tests, with attention focused on a handful of patent licensees including Myriad Genetics and Athena Diagnostics.

While these cases, especially Myriad's exclusive license to the BRCA1 and BRCA2 genes, have been central to the debate over the legitimacy and potentially harmful impact of gene patenting on treating patients, microarrays are invoked as a case where gene patents could create an "anticommons" or "thicket" of insurmountable IP—a characterization that some experts endorse and others question.

This chapter will offer a general overview of the state of gene patents, with a primary focus on the USA, and then will discuss some of the arguments that relate

J. Petrone (✉)
BioArray News, GenomeWeb LLC, New York, NY, USA
e-mail: jpetrone@genomeweb.com

B. Jordan (ed.), *Microarrays in Diagnostics and Biomarker Development*,
DOI 10.1007/978-3-642-28203-4_13, © Springer-Verlag Berlin Heidelberg 2012

to microarrays while providing some background on litigation in the array industry. At the end of the chapter, proposed solutions to some of the issues raised will also be discussed.

13.2 An IP Landscape in Flux

At this time more than ever, the legitimacy of patenting genes is being called into question. Most of the questions surrounding that legitimacy spring from one case, Myriad Genetics' exclusive license to the BRCA1 and BRCA2 genes, but they are not restricted to this particular instance. Other cases, and even government reports, have raised new questions about the future of gene patenting. It is within this uncertain environment that the debates about the impact of gene patents on microarray technology are occurring.

Myriad Genetics is a Salt Lake City-based company founded by former University of Utah researchers. Based on federally funded research, Myriad attracted investment from Eli Lilly and patented the BRCA genes and has offered since a test for breast and ovarian cancer called BRAC Analysis. While Myriad has mostly not enforced its rights in the case of researchers, it has sent out cease-and-desist letters to any laboratories, commercial and public, that offer tests analyzing patient's DNA for certain BRCA1 and BRCA2 mutations correlated with increased likelihood of developing these two cancers. This led to the end of US-based testing for the mutations, outside of Myriad's offering, though international response to the threat of litigation has been varied, with some labs continuing to test for mutations in these genes (Carbone et al. 2010).

Long standing "resentment" toward the firm resulted in a 2009 American Civil Liberties Union-led suit against Myriad and the US Patent and Trademark Office, which resulted in a March 2010 US District Federal District Court ruling that isolated DNA is not patentable in the USA and that Myriad's claims relevant to testing for the BRCA1 and BRCA2 genes were invalid (Carbone et al. 2010). However, a year and a half later, a US Appellate Court reversed the district court's ruling, deciding instead that genes were in fact patentable. The plaintiffs are now considering appealing the case to the US Supreme Court (Pollack 2011).

While the Myriad case has arguably captured the most attention, it has not occurred in a bubble. The firm is one of a number of IP holders that vigorously protect their portfolios. Others, such as Worcester, MA-based Athena Diagnostics and New Haven, CT-based PGxHealth have sought to stop university laboratories from offering genetic testing for diseases (Carbone et al. 2010).

A number of other cases and reports have also shed uncertainty on the legitimacy of patenting genes. In February 2010, the Secretary's Advisory Committee on Genetics, Health, and Society issued a draft report on the issue of gene patenting and its impact on diagnostics, where it found "no convincing evidence that patents either facilitate or accelerate the development and accessibility" of genetic tests and "some, albeit limited, evidence that patents had a negative effect on clinical research and on the accessibility of genetic tests to patients" (Carbone et al. 2010).

Some experts believe that the Myriad case and the SACGHS report reflect "growing disenchantment with the patenting and licensing practices of universities and industry." Moreover, they argue that the increased clinical use of microarrays and next-generation sequencing "deepens concerns" about gene patenting, and some argue that the "legacy of exclusively licensed gene patents casts a shadow of patent infringement liability over the future of multi-allele testing and full-genome analysis" (Carbone et al. 2010).

13.3 The Microarray "Anticommons"

The 2010 SACGHS draft report explicitly stated that microarray-based multiplex testing could be hindered by gene patents. The probe molecules used to detect gene sequences would infringe corresponding patented genes if the probe molecules are identical or equivalent to the claimed isolated genes, the authors of the report noted. Furthermore, the probes on the array would infringe any claims to identical or equivalent oligonucleotide molecules useful as probes. Multiplex testing would also infringe association patent claims, they argued in the report (SACGHS 2010).

Developers of array-based tests would need to obtain licenses to all patents infringed by molecules used in the test, the authors argued, noting that the alternative of leaving patented genes out of a test or not reporting the results related to those genes "undermines the very clinical utility of multiplex analysis." At the same time, obtaining such licenses could prove problematic, compelling the test developer to conclude agreements with many IP holders. The cost of obtaining these patents might make the test maker's product unprofitable or discourage them from developing multiplex tests in the first place, the authors note.

Another issue is determining whether a product or service could infringe existing patents. "Choosing to proceed with a product involves the risk of being sued, and the expense of defending against suits that arise diverts funds that could otherwise be used for innovation," the authors of the report added.

One frequently cited study (Jensen and Murray 2005) determined that roughly 20% of the genes identified in the human genome are referenced in the claims of patents, corresponding to 4,382 genes out of the 23,688 genes in the National Center for Biotechnology Information's gene database as of 2007. The genes references are owned by more than 1,000 assignees, more than 60% of which are private companies.

One caveat, according to SACGHS, is that even when a patent claim contains a nucleotide sequence, it does not necessarily mean that the isolated nucleic acid molecule that corresponds to that sequence is the patented invention. In some cases, the patent may be claiming the isolated molecule as the invention, but in other cases, the patent could be claiming something else, such as a process for using the molecule, the authors of the report wrote.

Still, the SACGHS report used the term "patent thicket" to describe the "dense web of overlapping intellectual property rights that a company must hack its way through in order to actually commercialize new technology." Array companies and users, the authors argued, face the "greatest level of complexity" in bringing a new test or tool to market. The authors also noted that vendors Affymetrix and Illumina have "raised concerns in public about patents affecting the development of multiplex tests." In its public comment on the draft form of the report, Illumina "expressed support for gene patenting, but pointed out that dealing with such vast amounts of genetic information has the potential to raise a whole host of unique intellectual property challenges."

"Patent thickets" are also described in literature as the "tragedy of the anticommons." In an "anticommons," the report notes, multiple owners "endowed with the right to exclude others from a scarce resource, and no one has an effective privilege of use. When there are too many owners holding rights of exclusion, the resource is prone to underuse—a tragedy of the anticommons." In a "microarray anticommons," a single multimarker assay that uses multiple biomarkers that are patented and licensed to multiple competing parties may never be commercialized because of the IP issues facing its developers (Park 2010).

Barton calls this issue "royalty stacking," making similar arguments. As each holder of a patent on a diagnostic sequence or marker used in the array could, under traditional law, legally block marketing or use of the array, the cost of those agreements may be significant and, when added to the financial burden of regulation, may slow the development of products (Barton 2006). He also argues that next-generation sequencing platform users could suffer from the royalty stacking problem, noting that "even one who uses a sequencing machine could infringe patents on using the information derived from any specific mutation found in the sequence."

13.4 "Harms of Exclusivity"

An April 2010 issue of *Genetics in Medicine* featured a discussion of the impact of gene patents on medical genetics, with papers pertaining to the way in which gene patents have influenced testing for a given disease. Two of the papers and a commentary explicitly mentioned array technology and made arguments similar to those found in the SACGHS report.

Discussing Athena Diagnostics' ownership of genes associated with risk of developing superior cerebellar artery syndromes (Powell et al. 2010), it was argued that enforcement of the patents could "reduce incentives to develop a chip-based or microarray-bead or sequence-based test using alternative technologies because the patents apply to any technology for assessing patented mutations or diagnostic methods that entail sequencing or sampling a patented sequence." In another study concerning hearing loss (Chandrasekharan et al. 2010), the authors noted that while array-based genetic testing for hearing loss was being developed,

Athena's ownership of patents covering genes related to hearing loss made it difficult for tests developed outside the USA, in countries where the patents are not enforceable, to be marketed to US customers. Asper Biotech, a Tartu, Estonia-based firm that offers microarray-based hearing loss testing using some of the genes owned by Athena, was cited as such an example. While Asper's tests are used in the USA as part of clinical research studies, it is the firm's opinion that if Asper was to market its test in the USA, it would have to acquire licenses to all relevant IP and factor royalty payments into its business plan. However, the author expresses some doubt over this scenario. "We are not sure we concur with this judgment if the test itself were conducted in Estonia," she writes.

A commentary in the same issue states that the case studies and the SACGHS report demonstrate a number of harms that result from gene patents in the diagnostic arena (Evans 2010). These "harms of exclusivity" include patient access to testing, the inability to obtain second-opinion testing, and concerns over test quality. "When only a single laboratory is allowed to engage in diagnostic testing, clinicians who are concerned about a particular laboratory have no recourse in their choice of laboratories, and it is the laboratory, rather than patients and providers, who define the terms of testing," the author notes.

13.5 The "Overstated" Impact on Arrays

The microarray industry has been marked by a series of lawsuits over the years. Most of the suits evolved out of IP awarded to a handful of technology developers who filed patent applications in the early 1990s, among them Hyseq, Affymetrix, Oxford Gene Technology, and Stanford University (Rouse and Hardiman 2003).

In 1997, Hyseq filed an infringement suit against Affymetrix, which ended in a 2001 settlement. OGT similarly initiated litigation against Affymetrix in 1999 which ended in Affymetrix obtaining a license to OGT's IP in 2001. Subsequent OGT lawsuits against a number of players yielded similar results. Another infringement case between Incyte Genomics, which had acquired licenses to Stanford's array-related patents, and Affymetrix also ended in a 2001 settlement.

More recent infringement suits have occurred between Affymetrix and Illumina, with a series of cases filed between 2004 and 2010 in both the USA and the EU. In the first series of suits brought by Affymetrix against Illumina, the latter firm chose to settle in 2007. A second round of suits brought by Illumina against Affymetrix was dismissed by both a district and a federal appellate court in 2010 and 2011.

Almost all of the array-related lawsuits filed to date have concerned technology IP, and according to one expert, no suit has ever focused on the infringement of a particular gene patent. Because of this, Holman has criticized opponents of gene patents of mistakenly assuming that a gene is patented because it is referenced in a patent claim. Instead, he maintains that the "vast majority of uses of the genes are not covered by the claims in these patents." Because of this, he argues that it is a "gross exaggeration to contend that Jensen and Murray's results indicate that 20%

of human genes are locked up by patents," but notes that "this assertion is made all the time" (Holman 2010).

According to Holman, the fears with respect to arrays and so-called patent thickets have "failed to materialize." As evidence, he uses Affymetrix as an example of a company that has sold microarrays since the early 1990s and not faced litigation related to infringing on a specific gene patent. Holman also conducted a review of human gene patent litigation in 2007 and argues that arrays have "never been the subject of a patent infringement lawsuit involving a gene patent."

Holman claims that the argument that patent thickets would block the development and use of arrays was based on two assumptions. One was that once a gene has been patented, no one can use it for any purpose and the other is that patents will always be enforced. As arrays often contain a short fragment of a gene, not the full sequence, gene patents that only claim the full-length sequences would not be infringed upon by a given array.

Another issue is how well such patents would stand up under legal scrutiny. Or, even if a patent covers a small number of the probes on an array, a royalty rate could be low, not justifying a patent infringement suit to begin with. Even if an injunction were to be entered, array companies could simply remove the gene fragment from the chip in question. In many cases, though, it would be possible to avoid the injunction by removing the infringing fragment from the array, as "the loss of a single gene from a microarray representing thousands of genes will in most instances not substantially impair the clinical utility of the array, and other genes would be available as substitutes if necessary" (Holman 2010).

As Holman has noted elsewhere, as few human gene patents have ever been asserted in court (his 2007 review identified 31 of such cases going back to 1987), any "chilling effect" of gene patents on array technology arises from a "perception of risk that may not comport with reality." A patent generally has no legal effect until successfully asserted in court, and attempts at judicial enforcement often fail, Holman has argued. As evidence, he stated that not one of the gene patents referenced by Jensen and Murray when tried resulted in a decision favoring the patent holder (Holman 2008). Ultimately, he holds that there is "little empirical support for a legislative bar to the patenting of genes or DNA."

13.6 Different Solutions

Though experts disagree on the extent to which gene patents obstruct the development of array-based tools and tests, there seems to be some consensus that some kind of reform is needed, and there is no shortage of proposals for ways to fix the system.

Holman has argued that if any legislative reform is necessary, it should be limited to an exemption from infringement liability for research or diagnostic testing uses of naturally occurring genetic sequences (Holman 2008). Such an approach "preserves the patentability of gene-based innovations, while curtailing some of the more problematic enforcement activities," he has stated.

Barton believes that reform could also come from funding agencies. As it is likely that many of such patents will derive from research carried out under public grant, he has proposed that the US National Institutes of Health impose on genetic research a self-denying injunction about patents that would effectively require licenses. In terms of legal reform, he argues that patent law could recognize that patents on genomic markers reflect a "different kind of innovation" and that in the future patent coverage for genes could be rejected or patent holders should be obligated to grant "reasonable royalty licenses for use of the sequence as part of an array or for use of the sequence in pharmacogenetics" (Barton 2006).

In its report, the SACGHS authors argued as much, requesting the Department of Commerce to promote several statutory changes, including the creation of an exemption from liability for infringement of patent claims on genes for anyone making, using, ordering, offering for sale, or selling a test developed under the patent for patient care purposes and the creation of an exemption from patent infringement liability for those who use patent-protected genes in the pursuit of research. If enacted, the first change would allow service providers to offer gene-based diagnostic testing "unimpeded by fear of infringing patent claims on genes and would apply to both commercial and noncommercial laboratories," the authors of the report hold. This would solve patient access problems that result from exclusive licenses to gene patents, they argue. In the report, the authors also urged the US Secretary of Commerce to discourage the "seeking, the granting, and the invoking of any patents on simple associations between a genotype and a phenotype." Association patent claims "threaten the availability of existing genetic tests and are an anticipated barrier to the development of testing innovations, such as microarrays and whole-genome sequencing," the authors wrote.

Carbone similarly argues for a "clear research exemption" in the case of gene patents, as well as a simplified means of challenging patents, and improved examination procedures to avoid overly broad patent claims. "Overall, the problem does not lie wholly in patent law but rather concerns how decisions are made about what is patented ... and how patents are managed and used," she argues. She also speculates that "with one or a few successful challenges to broad patents enforced for diagnostic purposes, the business models of enforcing monopolies on genetic testing for specific conditions would probably give way to more cross-licensing, more competition and faster innovation in testing methods."

Until such changes are made though, Park argues that researchers and developers of clinical multimarker assays should be prepared to deal head-on with the issues of biomarker and gene patents. Noting that arrays have been established as research tools, he points out the existence of several clinical microarray-based assays available in the European and US markets. "As technical issues of quality control are solved, these complex assays will inevitably become more clinically useful and commonplace; however, the large number of analytes that power these assays is also the potential Achilles' heel in terms of commercialization," he writes. "By embracing issues of intellectual property in the earliest stages of developing a multimarker assay, perhaps a microarray anticommons can be avoided."

References

Barton J (2006) Emerging patent issues in genomic diagnostics. Nat Biotechnol 24(8):939–941

Carbone J et al (2010) DNA patents and diagnostics: not a pretty picture. Nat Biotechnol 28(8):784–791

Chandrasekharan S et al (2010) Impact of gene patents and licensing practices on access to genetic testing for hearing loss. Genet Med 12(4 Suppl):S171–S193

Evans J (2010) Putting patients before patents. Genet Med 12(4 Suppl):S3–S4

Gene patents and licensing practices and their impact on patient access to genetic tests. Report of the secretary's advisory committee on genetics, health, and society. http://oba.od.nih.gov/oba/sacghs/reports/SACGHS_patents_report_2010.pdf. Accessed Apr 2010

Holman C (2008) Trends in human gene patent litigation. Science 322(5899):198–199

Holman C (2010) Gene patents under fire: weighing the costs and benefits. Biotechnology and software patent law: a comparative review of new developments. In: New directions in patent law. Edward Elgar, Cheltenham

Jensen K, Murray F (2005) Intellectual property. Enhanced: intellectual property landscape of the human genome. Science 310(5746):239–240

Park J (2010) The tragedy of the microarray anticommons. Clin Chem 56(11):1683–1685

Pollack A (2011) Despite gene patent victory, myriad genetics faces challenges. New York Times, Aug. 24

Powell A et al (2010) Spinocerebellar ataxia: patient and health professional perspectives on whether and how patents affect access to clinical genetic testing. Genet Med 12(4):S83–S110

Rouse R, Hardiman G (2003) Microarray technology–an intellectual property retrospective. Pharmacogenomics 4(5):623–632, Review

Concluding Remarks

I believe this book has provided readers with an extensive and essentially complete overview of microarrays in diagnostics and biomarker development, as announced by its title. Of course, some aspects have been left aside, such as living cell arrays (used in particular for siRNA screening), or microfluidics, since they are essentially used in research for the former, and constitute a whole field worthy of a special book for the latter. It is quite obvious from this overview that the field is very much alive, in contrast to a fairly common feeling that arrays are an obsolescent technology. Indeed, their development has not followed the path predicted by some analysts a decade ago, when e.g. Frost and Sullivan foresaw a total market of 3.3 billion USD by 2004 (BioArray News 2001), much higher than the current figure of less than one billion USD (in 2010). The recent and highly publicised progress in new-generation sequencing (NGS) has led to a widespread impression that arrays are on the way out, but this impression is definitely mistaken. NGS is indeed superseding arrays for a number of research applications such as expression profiling (RNA-seq) and epigenetic studies (Chip-seq). Nevertheless, arrays enjoy a secure position in other fields such as clinical CNV assessment as well as in many intermediate-plex genotyping or expression assays. Recent advances in protein arrays, as well as the development of peptide and aptamer arrays, bring the array paradigm to areas in which NGS is definitely not a competitor. In the widely announced era of personalised medicine (that is indeed coming of age, if more slowly than many predicted), molecular diagnostics will play an essential role, as the medical paradigm moves towards more and more closely tailored treatments based on assessment of genetic or metabolic characters that will in most cases involve multiplexed measurements. In this landscape, while arrays are not the "one-size-fits-all" solution touted 10 years ago, they definitely have their role to play in a wide range of biological fields, especially so in clinical practice where turnover time and cost issues assume great importance, and in protein or activity assays

B. Jordan (ed.), *Microarrays in Diagnostics and Biomarker Development*,
DOI 10.1007/978-3-642-28203-4, © Springer-Verlag Berlin Heidelberg 2012

where DNA or RNA cannot provide the required information. It may well be that another instalment of this book will become necessary a few years from now – I would expect it to be less based on DNA probes than the current volume, and to incorporate practical implementations of miniaturised, lab-on-chip and portable systems.

Marseille, December 2011 Bertrand Jordan

Jones MM (2001) Analysts predict red hot future for Biochips: sector poised to grow to $3.3 billion by 2004? BioArray News 1:2–3

Subject Index

B. Jordan (ed.), *Microarrays in Diagnostics and Biomarker Development*,
DOI 10.1007/978-3-642-28203-4, © Springer-Verlag Berlin Heidelberg 2012

L

M

N

O

P

R

S

U

V

MIX
Papier aus verantwortungsvollen Quellen
Paper from responsible sources
FSC® C105338

If you have any concerns about our products,
you can contact us on
ProductSafety@springernature.com

In case Publisher is established outside the EU,
the EU authorized representative is:
Springer Nature Customer Service Center GmbH
Europaplatz 3, 69115 Heidelberg, Germany

Printed by Libri Plureos GmbH
in Hamburg, Germany